Best Practices in Metropolitan Transportation Planning

Planning at a metropolitan scale is important for effective management of urban growth, transportation systems, air quality, and watershed and greenspaces. It is fundamental to efforts to promote social justice and equity. *Best Practices in Metropolitan Transportation Planning* shows how the most innovative metropolitan planning organizations (MPOs) in the United States are addressing these issues using their mandates to improve transportation networks while pursuing emerging sustainability goals at the same time.

As both a policy analysis and a practical how-to guide, this book presents cutting-edge original research on the role accessibility plays – and should play – in transportation planning, tracks how existing plans have sought to balance competing priorities using scenario planning and other strategies, assesses the results of various efforts to reduce automobile dependence in cities, and explains how to make planning documents more powerful and effective.

In highlighting the most innovative practices implemented by MPOs, regional planning councils, city and county planning departments and state departments of transportation, this book aims to influence other planning organizations, as well as influence federal and state policy discussions and legislation.

Reid Ewing is chair of the Department of City and Metropolitan Planning at the University of Utah, USA, associate editor of the *Journal of the American Planning Association*, and columnist for *Planning* magazine. Ewing's nine books and over 100 peer-reviewed articles are aimed at planning practitioners.

Keith Bartholomew is an Associate Dean and Associate Professor at the University of Utah College of Architecture + Planning, USA. Professor Bartholomew previously served as a staff attorney for 1000 Friends of Oregon and was the director of "Making the Land Use, Transportation, Air Quality Connection."

"Ewing and Bartholomew expertly draw together a rich, well-researched, and sophisticated study of metropolitan transportation planning. Along with their discussion of federal policy and regional plans, the quantitative analyses and best practices form a strong, evidence-based foundation that is both scholarly and practical. This is an excellent resource for practitioners, academics, and students interested in these important planning issues."

– **Tom Sanchez**, *Chair and Professor, Urban Affairs and Planning, Virginia Tech, USA*

"Ewing and Bartholomew effectively document conventional and best practices in metropolitan transportation planning – highlighting the need for more expansive and sophisticated approaches. They say, '. . . transportation professionals have been in somewhat of an existential crisis for a couple of decades.' So true, in large part because they did not have the tools nor the political support to advance more sustainable transportation. Now that the world is being turned topsy-turvy with the embrace of climate policies and the advent of a slew of innovative mobility services and technologies, we need a major overhaul of how we plan and manage transportation. This book is a good start."

– **Daniel Sperling**, *Professor of Transportation, University of California, Davis, USA, and Author of* Three Revolutions: Steering Automated, Shared, and Electric Vehicles to a Better Future

"Ewing and Bartholomew review the recent evolution of metropolitan transportation planning organizations though the quantitative and qualitative analysis of over 100 regional plans to understand both the analytical tools and substance of the planning process, highlighting the recent emphasis on regional-level scenario planning in federal legislation. This book is a great addition to libraries of planning researchers and good for practitioners who want to understand the role of regional transportation planning in creating a sustainable metropolis."

– **Ruth L. Steiner**, *Professor and Director of the Center for Health and the Built Environment, University of Florida, USA*

"This book provides an invaluable resource for transportation planners to learn about innovative approaches and best practices being utilized in other regions around the country. Metropolitan Planning Organizations should have this book within arm's reach."

– **Andrew Gruber**, *Executive Director, Wasatch Front Regional Council, Utah, USA*

Best Practices in Metropolitan Transportation Planning

Reid Ewing and Keith Bartholomew
with Alexander Barton, Allison Spain,
and Pratiti Tagore

LONDON AND NEW YORK

First published 2018
by Routledge
2 Park Square, Milton Park, Abingdon, Oxon OX14 4RN

and by Routledge
711 Third Avenue, New York, NY 10017

Routledge is an imprint of the Taylor & Francis Group, an informa business

British Library Cataloguing-in-Publication Data
A catalogue record for this book is available from the British Library

Library of Congress Cataloging-in-Publication Data
Names: Ewing, Reid H., author. | Bartholomew, Keith, author.
Title: Best practices in metropolitan transportation planning /
Reid Ewing and Keith Bartholomew.
Description: 1 Edition. | New York : Routledge, 2018. |
Includes bibliographical references.
Identifiers: LCCN 2018000660 | ISBN 9780815381006 (hardback) |
ISBN 9780815381037 (pbk.) | ISBN 9781351211345 (ebook)
Subjects: LCSH: Urban transportation policy—United States.
Classification: LCC HE308 .E95 2018 | DDC 388.40973—dc23
LC record available at https://lccn.loc.gov/2018000660

ISBN: 978-0-8153-8100-6 (hbk)
ISBN: 978-0-8153-8103-7 (pbk)
ISBN: 978-1-351-21134-5 (ebk)

Typeset in Sabon
by Apex CoVantage, LLC

Contents

Acknowledgments

Material in this book is based upon work supported by the National Science Foundation under Grant No. BCS – 1224102 "Measuring the Environmental Costs of Space-time Prisms in Sustainable Transportation Planning." Funding was also provided by the Federal Highway Administration under Purchase Order No. DTFH61–08-P-00191. Any opinions, findings, and conclusions or recommendations expressed in this publication are those of the authors and do not necessarily reflect the views of the funding agencies.

Notes on contributors

Keith Bartholomew is an Associate Dean and Associate Professor at the University of Utah College of Architecture + Planning. Professor Bartholomew previously served as a staff attorney for 1000 Friends of Oregon and was the director of "Making the Land Use, Transportation, Air Quality Connection" (LUTRAQ).

Alexander Barton is a master's student in the University of Utah's Department of City and Metropolitan Planning. He received a bachelor of geography degree from Brigham Young University, with an emphasis in urban and regional planning and a minor in psychology.

Reid Ewing is chair of the Department of City and Metropolitan Planning at the University of Utah, associate editor of the *Journal of the American Planning Association*, and columnist for *Planning* magazine. Ewing's nine books and over 100 peer-reviewed articles are aimed at planning practitioners.

Allison Spain received her master's from the University of Virginia in Urban and Environmental Planning. She served as staff for the Department of City and Metropolitan Planning at the University of Utah for two years, and currently works for the Virginia Food System Council, a nonprofit focused on strengthening Virginia's food system.

Pratiti Tagore is a PhD student in the Department of City and Metropolitan Planning at the University of Utah. She holds a bachelor's degree in architecture, and a master's degree in urban planning, both from India. Pratiti studies the influence of built and natural environments on people's perceptions, and how perceptions affect actions.

1 Introduction and context

Salt Lake City, the central municipality in the Salt Lake City region, was founded in 1847. The City of Sandy, Utah, 13 miles away at the southern end of the Salt Lake valley, was settled beginning in the 1860s. At the time of the latter's creation, the two jurisdictions were separated by orchards, open fields, and a lot of scrub. A single road – State Street – connected the two bergs (Figure 1.1). But travel between the two was limited. Some commerce traveled up and down the road – ore from nearby mines and produce from the community's farms, mainly – and people made occasional trips for church or social purposes. But basically, the two communities were self-contained. Salt Lakers tended to stay in Salt Lake and Sandyites stayed in Sandy.

Then, things changed. First came the railroad, with the Utah Southern connecting the two cities in 1871 (Tullidge, 1886). The interurban streetcar followed in 1907 and "carried a rising number of workers from Sandy to jobs downtown and shoppers in search of a greater variety of goods" (Bradley, 1993, p. 106). The automobile arrived in the Salt Lake valley in 1900 and started making a real presence in 1917 when vehicle registrations hit 21,576, nearly double the number from the previous year. By the onset of the Great Depression in 1930, the numbers reached more than 96,000 (FHWA, 1995).

After the Depression and World War II, interaction – economic and otherwise – between Salt Lake and Sandy increased dramatically. This, naturally, increased market attention on the lands between the two cities, with many of the farms and open lands being converted into commercial and residential developments. Annexations led to a 36 percent increase in Sandy's jurisdictional territory during the 1950s, and in 1969 the city doubled in size during a single year. The Interstate Highway System came to the region during this same period. Constructed during the late 1960s/early 1970s, Interstate 80 tied the valley together east and west, whereas I-15 connected it north to south. Other links – I-84 through Ogden and the I-215 circumferential beltway – filled out the region's interstate pallet.

With all of these connections between the two cities, people began to travel back and forth much more frequently. Daily commuting to work between Sandy and Salt Lake had become the rule, not the exception.

Figure 1.1 State Street, Salt Lake City, ca. 1907.
Used by permission, Utah State Historical Society.

And people traveled between the two cities for other purposes, as well – shopping, entertainment, education, recreation. By 1978, 65 percent of Sandyites reported that whereas they bought food locally, they traveled to other towns for other shopping (Bradley, 1993). Eventually, the rapid dispersion of employment to suburban places during the 1980s and 1990s led toward a partial balancing out in directional commute patterns, with people traveling to jobs in suburban towns like Sandy as much as they were traveling to Salt Lake City.

At some point during this history, the boundaries of each of the two cities – and all of the other jurisdictions between – became much less important to people's daily lives. Whereas in earlier times, people might go for days or weeks without crossing their home city's boundaries, now they were crossing those boundaries daily, frequently several times a day. In the last year of the 20th century, intra-urban passenger rail began returning to the region after a half-century hiatus. By 2013, approximately 150 miles of light rail, heavy rail, and streetcar service helped tie together a vast interconnected and integrated landscape of 1,600 square miles and 1.8 million souls.

Figure 1.2 State Street, Salt Lake City, 2017.
Photo: Keith Bartholomew.

The history of this one region mirrors the history of urban places all across America, and indeed across the globe. In the span of approximately one and a half centuries, our urban places have shifted from very localized nodes of semi-independent communities to expansive territories of interdependent conglomerations that function as webs of social and economic interaction. As Nelson and Lang note, "Cities are defined by their spatially integrated functions, not by their political boundaries. The suburbs and even exurbs of a city are elements of the city writ large" (2011, p. 1). In the shifts from mono-centric urban forms to poly-centric regions to "edgeless" cities, noted by Lang (2003), urban areas have become increasingly characterized by multidimensional overlapping networks of social, cultural, and economic ties that knit together people and institutions in ways that make only passing acknowledgment of municipal jurisdictional boundaries. Urban regions are the new "uber-networks."

Economically, metro regions are, overwhelmingly, the source of America's wealth. Katz and Bradley (2013) report that the nation's 100 largest metro areas generate 75 percent of U.S. GDP. This prosperity comes not from an aggregation of a multitude of municipally based individual economies, but from regions operating as integrated systems. Indeed, metro regions appear to be eclipsing other geographic units – including states and nation-states – in economic importance. Referencing Kenichi Ohmae's *The End of the Nation State*, Dan Kemmis notes that "more organic entities such as continents, city-regions, and coherent subcontinental regions are rapidly emerging as considerably more relevant economic entities than states, provinces, and nations" (2001, p. 73). This is leading to what some refer to as an inversion of the traditional top-down federal-state-local hierarchy to a new power structure that places metro areas at the top.

The importance of the regional unit is demonstrated even more dramatically through ecological imperatives. Nothing in nature respects the political boundaries created by humans. Wildlife, air and the things that pollute it, vegetation, and the movement of water all operate according to forces that are independent of jurisdiction and government (though governments frequently take massive actions to influence these forces). In fact, it was the inadequacy of state boundaries to address the trans-boundary nature of watersheds that was a primary motivation for convening the U.S. Constitutional Convention in 1787.

Regions hang together through social institutions as well. As Calthorpe and Fulton (2001) note, cultural organizations, such as opera and ballet companies, sports facilities, religious temples and cathedrals, universities, and hospitals all operate through and for region-level populations, not just those of the municipality in which they are located. And increasingly, these institutions are not only located in central cities. In the Salt Lake area, the arena for the region's NBA franchise is in Salt Lake City, but the Major League Soccer team is in Sandy, and the hockey team is in West Valley City. Together, institutions create webs of interactions that further cement the idea of a region-scale sense of place. We might reference our individual neighborhoods or suburban towns when speaking to someone else from the same region. But when we travel nationally or internationally, we tell people that we live in San Francisco, or Seattle, or Chicago, or Salt Lake City, meaning not that we necessarily live within those municipalities, but that we come from those metro regions. In other words, we conceptualize "home" as a regional thing. As David Rusk has observed, "the real city is the total metropolitan area – city and suburb" (1993, p. 5) – the two are inseparable in cultural as well as functional terms.

Despite the economic, ecological, and social strands that tie regions together, policy and governance structures in American regions remain largely the purview of state and municipal governments, units that are either too large or too small to reflect the lived realities of people in metropolitan areas. What is necessary are funding and policy structures that "capture the true economic and social geography" of multijurisdictional regions (Carbonell & Yaro, 2005). In other words, there needs to be governmental institutions and structures that have a geographic scope commensurate with the scale of how people live and environmental systems function.

The history of metropolitan regionalism in the United States and Europe is more than a century old, beginning with planning around the expanding capitals of northern Europe (Ward, 2002) and continuing with the pioneering work in the United States by the organization that would become the Regional Plan Association (Yaro & Hiss, 1996). Much of this work, however, has been championed by academicians and nonprofits, whereas governing institutions have largely remained impervious to regional approaches. As Seltzer and Carbonell (2011) outline, long-standing local government boundaries "take on cultural meaning associated with values" and as such are resistant to change.

Despite this inertia, there is one area in which a regional policy structure has gained traction: transportation. The inherent regional nature of transportation systems and the proverbial worry that highways and rail lines might not meet at municipal boundaries absent a regional approach has proven sufficiently compelling to overcome the institutional intransigence. At the federal level, these concerns culminated in Congress' passage in 1962 of legislation that, for the first time, required transportation planning to occur at an "urban" scale, that is, in a frame tied to the form of settlement patterns on the ground, not to jurisdictional boundaries. In the 1970s, Congress took a further step, requiring that this planning be done by "metropolitan planning organizations" (MPOs), conglomerations of urban governmental representatives. Although for a brief time, MPOs were tasked with other regional planning tasks, it has been transportation planning that has remained the bedrock of regional planning efforts in the United States.

In this book, we review that history, tracing the arc of planning for more than a half-century, highlighting what we perceive as the best practices currently being pursued by American planning agencies. Through this assessment, our objective is to address what it means to undertake regional transportation planning in America today and to advance those approaches that are leading America toward a better tomorrow.

We begin this assessment in Chapter 2 where we set basic analytic principles for the meaning of cities and the purpose of transportation. Understanding that planning is meaningless without a definition of purpose and a reflection on motivations, we take a step back from much of the planning practice of the 20th century to explore the real function that both cities and transportation play.

Chapter 3 continues with a quantitative analysis of current regional transportation plans now in place in 110 U.S. regions. This snapshot gives an overview of what our current generation of plans is yielding in terms of transportation investments and regional outcomes.

In Chapter 4, we shift the focus to zoom in on integrated land use-transportation scenario analysis, an approach to regional transportation planning that emerged in the 1990s and has gained significant momentum in the 21st century.

Chapter 5 examines the net results of transportation planning since World War II by looking at the historical growth in transportation consumption over the decades. Although these trends are driven by more than just what planners provide, our current understanding of induced-demand effects on capacity/demand relationships informs us that the plans that planners develop, and the facilities that get built according to those plans, have a significant role to play in the transportation behavioral patterns we see today in American metro areas.

Chapter 6 continues our investigation with a deep-dive qualitative investigation into the top transportation plans in the country. Through this treatment, we highlight those policies and practices that are truly at the practice's cutting edge.

Chapter 7 closes the inquiry with a reflection on where we see the practice of metropolitan transportation planning headed in coming years. Among other topics, we tackle one of the persistent challenges for planners: how to measure and operationalize accessibility so that it can (finally) enter the mainstream of planning practice.

Today, we live in a very different world from that of the 1950s, when gas was 20 cents a gallon and President Eisenhower launched the interstate highway system. Urban sprawl is the nation's dominant development pattern. Americans are stuck with costly commutes and congestion. Roads and bridges are poorly maintained. Our climate is threatened and long-run oil supplies are uncertain. Massive demographic shifts suggest the need for alternatives to the automobile. The nation needs a transportation system that is ready for the rapidly changing economy of the 21st century. We believe that the planning practices we observe in this volume provide the beginning of a new approach to regional transportation planning.

2 Accessibility and the purpose of cities

> [The city] comes to be for the sake of living, but it remains in existence for the sake of living well.
>
> (Aristotle)

Transportation systems provide the armature around which neighborhoods, towns, and regions are constructed, and the nature of those systems significantly influences the quality of the surrounding communities. Conversely, the nature of a community can significantly influence the functioning of the transportation systems that serve that community. In observing the interactions between transportation and community, it is important first to note that transportation is primarily a derived demand – most of its value is in the role it plays in facilitating the functions of daily life, not in the actual activity of moving (Hanson, 2004; Mokhtarian, Salomon, & Redmond, 2001).

That being so, it would seem that an obvious question for transportation planners would be: How can all the activities that go into making a full and meaningful life be accomplished without excessive travel? Given the enormous level of environmental and fiscal resources needed to build, maintain, and operate community and regional transportation systems, it would seem logical that a primary role for transportation planning would be to facilitate daily household activity needs while minimizing the transportation costs of the individual and society. Yet, despite the occasional hortatory reference, this calculus has rarely played a central role in transportation planning processes. Rather, the predominant concerns have been how to reduce vehicle congestion during peak periods and how to increase vehicle travel flow at identified traffic bottlenecks (Olson, 2000).

If transportation systems and the communities that are both served and affected by those systems are to be fully integrated in a planning process, the focus of planning practice needs to shift from mobility to accessibility. Mobility, which has been the primary focus of U.S. transportation planning for more than a half-century, focuses on the movement of vehicles in time and space. Accessibility, on the other hand, focuses on the ease and convenience with which a person (or an increment of freight) can gain access to a needed activity (Miller, 2005). Whereas the two concepts are related, they are not

Photo: Keith Bartholomew.

the same. In many circumstances increased mobility can increase a person's level of accessibility to daily destinations. Assuming a constant geographic arrangement of destinations and a low level of traffic congestion, the number of places one can access in a car is frequently much greater than the number accessible by walking. Accessibility, however, depends on land use patterns as well as mobility. For example, a person's access to grocery shopping can be increased by the construction of a new store within walking distance of that person's home or place of work. This would result in a net increase in the person's accessibility in a way that is independent of mobility. Accessibility can also be increased through the use of information technologies, such as the Internet, in ways that do not have a direct geographic component.

Striving for ever-increasing levels of mobility can, in fact, result in reduced accessibility. Increasing mobility by increasing transportation system capacity has facilitated a steadily expanding development pattern (Newman & Kenworthy, 1999). The saturation of automobile-related infrastructure and corresponding automobile dependency within dominant segments of American society have helped lead to a dispersion of destinations, thereby increasing the distances between where people are and where they need to go. In a self-fulfilling way, this has made access to many destinations increasingly dependent on high levels of mobility, frequently achievable only by automobile (Hanson, 2004). These mobility levels, however, are not shared across society. According to the U.S. Census, in 2010 more than 65.7 million

Americans (21.3% of the population) were too young to drive (age 15 or younger); more than 27.8 million (9%) were 70 or older and may have limited or reduced ability to drive; more than 37.1 million (12.1%) reported a physical disability, which may impair driving abilities; and more than 46.6 million (15.4%) had income below the poverty level, making the ownership and operation of an automobile financially prohibitive. Moreover, maintaining and increasing mobility is becoming difficult to accomplish fiscally and environmentally. A 2006 study estimates that relieving severe traffic congestion in the United States would require an additional 104,000 lane-miles of capacity at a cost of $533 billion (Hartgen & Fields, 2006).

Even within the car owning and driving portions of society, a narrow focus on planning for transportation infrastructure that increases vehicle speeds may actually make it more difficult to achieve mobility goals by fueling a "vicious cycle" of induced demand that in turn influences land use and subsequent travel behavior (Silva, 2013). The vicious cycle works something like this: In an effort to combat congestion, roads are expanded or added, which initially reduces travel times and increases mobility. In many cases, these new roads "unlock" land at the urban periphery for development. Because people can travel farther in a given time, there is (at first) no accessibility penalty for firms, employers, and housing consumers to pursue inexpensive land at a city's periphery. This encourages destinations to spread farther apart, increasing automobile dependency (Ewing, 2008; Waddell, Ulfarsson, Franklin, & Lobb, 2007). Over time, new roads gradually fill up with traffic again as more people take advantage of greater mobility, a phenomenon known as induced demand (Goldman & Gorham, 2006; Gorham, 2009). As traffic and development demand gradually increase, Euclidean zoning and other land use policies designed to reduce congestion in developed areas by limiting density actually push development (and traffic) further out to the urban fringe, increasing both travel distances and times, as well as automobile dependency (Ferreira, 2012).

The cruel irony is that mobility-focused transportation policy often results in less mobility (Simpson, 2002). This happens because mobility-motivated transportation "improvements" can have significant impacts on land use patterns and the built environment, but conventional mobility metrics do not – and cannot – take these interactions into account. Instead of focusing on the need to ease or increase the flow of vehicles, transportation planning should focus on facilitating accessibility to needed and desired activities. In other words, the object of transportation planning should shift from looking at the needs of the machine to the needs of the operator.

The purpose of cities

The fundamental idea behind cities is permanence: permanent shelter, permanent storage, and permanent places of commerce (Mumford, 1938). The emergence of cities is, hence, tied with cultural shifts from nomadic hunting/

gathering to cultivated agriculture (Morris, 1996). Once groups of humans began putting down literal botanical roots, they began putting down social roots as well.

At first, becoming rooted in one location allowed only small expansions in the basic social unit: from the family to the clan (Diamond, 1999). Going from the clan settlement to a real "city," however, required two further advances: an agricultural surplus and some form of writing. With these developments, a nonagricultural class of labor could be created, and settlements could go beyond being just a collection of farmers and begin to achieve the kinds of complexity associated with urban development (Sjoberg, 1965). This change required technological advances that could facilitate the production of surplus food. The advances came together in the Tigris/Euphrates portion of the Fertile Crescent about 9,000 years ago. The technical innovations, and the food surpluses they facilitated, gave rise to whole categories of livelihoods that were only indirectly related to agriculture. These occupations, in turn, spawned still others; complexity was, in a sense, breeding itself.

By the time of the creation of Greek and Roman cities, the concept of urban settlement included, almost universally, the functions of marketplace, government administration, judicial adjudication, entertainment, social gathering, industry, and religious observance (Sperber, 1998). Many had notice boards for the posting of proclamations and other pieces of public information; pubs or drinking houses, which facilitated prostitution as well as social exchange; sports arenas; harbors and shipyards; and bathhouses, which were also a place of social gathering in addition to providing for personal hygiene. It is little wonder, then, that the Greek word *polis* has been interpreted to have at least seven distinct meanings in modern English, including stronghold, city, territory, social community, body of government, and political community (Hansen, 1998).

Over the centuries, cities became the central locations of culture, commerce, religion, education, and politics. These were the places of the concert hall, the temple, the market, the academy, and the castle. In cities, people sought profit, solace, security, knowledge, entertainment, power, and, somewhat ironically, solitude. Referring specifically to New York City, E. B. White (1949, p. 19) once observed that the city "is the concentrate of art and commerce and sport and religion and entertainment and finance, bringing to a single compact arena the gladiator, the evangelist, the promoter, the actor, the trader and the merchant."

In a word, people come to cities seeking exchange. Not just the exchange of the market – although that is of central importance – but also exchanges of ideas, cultures, values, and (not infrequently) DNA. Quoting Mumford (1938, pp. 5–6): "Through its concrete, visible command over space the city lends itself, not only to the practical office of production, but to the daily communion of its citizens." The city, in fact, "function[s] as the specialized organ of social transmission" (Brandford & Geddes, 1919, p. 156).

Virtually all of the institutions that are central to our current society have their roots in the exchanges that are facilitated by the proximity of humans

Photo: Keith Bartholomew.

to each other found in cities and towns (Bellah, Madsen, Sullivan, Swidler, & Tipton, 1992). Indeed, cities have become synonymous with civilization and culture (Hall, 1998), and even with life itself:

> Until lately the best thing that I was able to think of in favor of civilization, apart from blind acceptance of the order of the universe, was that it made possible the artist, the poet, the philosopher, and the man of science. But I think that is not the greatest thing. Now I believe that the greatest thing is a matter that comes directly home to us all. When it is said that we are too much occupied with the means of living to live, I answer that the chief worth of civilization is just that it makes the means of living more complex; that it calls for great and combined intellectual efforts, instead of simple, uncoordinated ones, in order that the crowd may be fed and clothed and housed and moved from place to place. Because more complex and intense intellectual efforts mean a fuller and richer life. They mean more life. Life is an end in itself, and the only question as to whether it is worth living is whether you have enough of it.
>
> (Oliver Wendell Holmes, March 7, 1900 (2004, pp. 85–86))

What is it about cities that facilitates such exchange? Among other things, it is because those exchanges can occur with relative ease and frequency, features that are facilitated by having concentrations of people near each other geographically. It is, in a very real sense, a matter of transportation. Recalling that the first real cities depended upon a surplus of food that could be exchanged for nonagricultural goods and services, those exchanges could only occur through a certain amount of transportation: Both the farmer and the tradesperson had to get their respective offerings to the point of exchange and home again. Hence, transportation became one of the early transaction costs underlying the emerging market system (Rao, 2006). As with other nonproductive costs, emphasis was placed on being able to keep transportation costs as low as possible. Cities, as it turns out, were adept at keeping transportation costs relatively low: The closer in geographic proximity the producer and consumer, the less travel is needed to consummate a transaction (Niewiem, 2005).

Cities facilitate frequent exchanges because in cities those exchanges can occur with a minimum amount of transportation: The locus of human interaction – be it in the marketplace, the local pub, the university, or the place of worship – is close and convenient enough for many to have ready and repeated access to it. The closer the place of exchange is, the less travel is required to access it, making it more convenient and resulting in increased levels of exchange opportunities. Hence, the level of exchanges and the amount of travel required to achieve them appear to have an inverse relationship: Beyond a certain minimum level of movement, the level of exchange seems to decrease as the amount of travel (measured in time or distance) increases.

As author/activist David Engwicht (1993) aptly observes: "[C]ities are an invention to maximize exchange (culture, goods, friendship, knowledge) and minimize travel." To put it into the language of mobility and accessibility, what is needed is an urban environment that is access efficient – one that provides the greatest amount of access to the places people need to go with the least amount of mobility.

From this perspective, it would appear that the primary function of transportation planning would be to reduce the amount of travel needed to accomplish the exchanges that make for a meaningful life. Traditional transportation planning processes, however, have not focused on exchange, but on the trips in-between exchanges, almost as if the trips themselves were the whole point. Although the concept of accessibility is not new – William Hansen, an engineer for the U.S. Bureau of Public Roads, first proposed the use of accessibility measures in transportation planning in the 1950s (Hansen, 1959) – a professional consensus about the best way to apply accessibility measures in practice has only emerged during the past decade (Halden, 2011).

A number of reasons account for this gap. Until the advent of computer-based geographic information systems, the difficulty of gathering sufficient data on a large number of origins and destinations, as well as the computing power needed to calculate metrics, was an obstacle (Anderson, Levinson, & Parthasarathi, 2013). Similarly, institutional inertia and the challenges of cross-disciplinary communication meant that accessibility has not always been "translated" into performance measures that direct planning efforts (Páez, Scott, & Morency, 2012, p. 141). Part of the problem is that land use and transportation practitioners often have different backgrounds and speak different professional languages. Unfortunately, the result of this split is that interpretable measures integrating sustainable accessibility to policy and practice are relatively scarce.

The lack of leadership from federal policy makers has also played a role. Whereas federal statutes and regulations have long listed mobility as a top priority, accessibility has been treated as a secondary concern, if addressed at all.

Federal law and transportation systems planning

Early efforts

At least at a project level, transportation planning has been occurring in the United States for a couple of centuries: the early turnpike roads built in Virginia and Pennsylvania in the late 18th century (FHWA, 1976) and the Transcontinental Railroad in the middle of the 19th (Bain, 1999) required a great deal of planning and engineering. The first significant systematic planning effort, however, began as the result of the Federal-Aid Highway Act of 1934 with a series of state-wide highway planning surveys done cooperatively by the U.S. Bureau of Public Roads (precursor to the Federal Highway Administration (FHWA)) and individual state highway departments (Weiner,

1999). These surveys mapped out the extent and physical characteristics of the highways within each state, and made some effort to quantify the volume of traffic on the various facilities. The cumulative result of these cooperative federal-state efforts was the beginning of a rudimentary nationwide highway plan. This initiative was soon dwarfed by the planning and engineering undertaken to construct the Interstate Highway System, which was authorized by Congress in 1944 but did not receive significant funding until the Federal-Aid Highway Act of 1956.

Whereas these efforts were distinct from their project-oriented predecessors in that they focused on entire systems, the focus was not on the daily travel needs of Americans going to and from their daily tasks. Rather, with respect to the Interstate Highway System, the objective was to create a system "which best and most directly join[s] region with region and major city with major city" (National Interregional Highway Committee, 1944). Moreover, the planning completely ignored important nonhighway transportation systems such as waterways, local arterials, and public transit. More to the point, all these planning efforts were pursued to construct facilities that would accommodate mobility; there was little to no interest in examining ways to reduce the need to travel: "The ultimate result of highway planning . . . is the development of a highway system to provide for the movement of vehicles" (FHWA, 1976).

The real shift in transportation planning came with the Federal-Aid Highway Act of 1962. Under this Act, metropolitan areas – defined as "urban area[s] of more than fifty thousand population" – were required to adopt long-range transportation plans for entire urban areas and for multiple modes of transportation. Rather than sporadic planning for a particular project (or even a specific system, such as the Interstate Highway System), the planning required under the Act was to be multimodal and "continuing, comprehensive, and cooperative" (the "3Cs") – meaning that planning had to be ongoing, not just a single event, incorporating a broad range of subjects and values, and carried out with the cooperation of state and local government agencies.

Just as important was the systematic incorporation of the effects that land use patterns have on travel demand. Although several significant transportation studies pioneered "hyphenated" land use-transportation analyses in the 1950s (Weiner, 1999), the incorporation of land use influences into transportation planning was routinized through the 1962 Act and subsequent implementing technical memoranda and procedural manuals developed by the Bureau of Public Roads (1963). In addition, there was a growing recognition of the community and social impacts that highway construction was having on urban areas, and a broader understanding of the need for highways to be better integrated with their surroundings. Despite these fundamental changes, however, the primary purpose of transportation planning remained focused on mobility and on ways to accommodate growing travel, primarily by automobile (Solof, 1997).

Photo: Keith Bartholomew.

The modern practice of transportation systems planning came of age in the decades following passage of the 1962 Act. That practice was focused on the development of multimodal transportation systems; it was facilitated and quantified by increasingly robust computer models; and it was executed by a new governmental entity created to meet the needs of the 3C planning process: the metropolitan planning organization (MPO) (Federal-Aid Highway Act of 1973). The rise of MPOs in the early 1970s grew out of a combination of increased environmental awareness, mounting resistance to urban highway construction, and heightened concern over the financial health of transit companies recently taken over by cities and other governmental entities. Prior to the formal recognition of MPOs in federal law, the regional intergovernmental bodies that facilitated the 3C process were playing, at most, technical and advisory roles: "This left the crucial day-to-day decisions about allocating funding and choosing projects largely to highway-oriented state officials" (Solof, 1997). In a fractious debate over reauthorization of federal transportation legislation in 1972–1973, a combination of urban, community, and environmental constituencies pushed to have these ad hoc regional bodies formally recognized in federal law, universalized to all metropolitan areas, and empowered to have decision-making authority. Once MPOs were formally recognized, as many as 39 other federal programs requiring regional planning came within their purview, in addition to transportation (Lewis & Sprague,

1997; McDowell, 1984). This golden era of MPOs, however, was short-lived. With the "new federalism" promoted by the Reagan administration, MPOs lost most of the programs they briefly controlled. The one program remaining was transportation planning, but new regulations gave states full sway in determining the functions for MPOs. This meant that many MPOs were in the role of merely rubber-stamping decisions already made by state highway departments (Solof, 1997).

Clean Air Act

Although important institutional changes occurred in the 1970s and 1980s that directly impacted transportation planning, most of the policy and procedural changes that occurred during those decades were imposed by broader environmental legislation that affected project planning (i.e., how/whether to construct a specific project), not systems planning. The first major policy directive specifically targeted at transportation systems planning since the 3C provisions in 1962 came as part of the Clean Air Act (CAA) Amendments of 1990. Substantial progress had been made in cleaning up the nation's air since the passage of the original CAA in 1970. Nevertheless, in 1990, 96 metropolitan areas still failed to meet national ambient air quality standards (NAAQS) for ozone and 41 failed to meet NAAQS for carbon monoxide, two pollutants associated with transportation sources. To remedy these failings, the CAA Amendments strengthened two components from earlier versions of the CAA, potentially turning each from being minor check-offs into major influences in transportation system planning processes.

The first was part of a new sliding scale compliance schedule for metropolitan areas out of attainment with the NAAQS: Those areas classified as "severe" or "extreme" for ozone, or "serious" for carbon monoxide had to adopt transportation control measures (TCMs) to offset projected growth in vehicle travel over the course of the planning horizon. Metropolitan areas could select from a long list of TCMs enumerated in the CAA Amendments, including programs to improve public transit and other alternative transportation modes, restrict automobile use in downtown areas, and "generally reduce the need for single-occupant vehicle travel, as part of transportation planning and development efforts of a locality, including programs and ordinances applicable to new shopping centers, special events, and other centers of vehicle activity" (42 U.S.C. sec. 7408(f)(1)(A)(xiv)). This option clearly lends itself to an access efficient outcome – facilitating access to needed destinations while working to reduce vehicle travel. The literature, however, suggests that few nonattainment areas have opted for this choice.

The second and further-reaching change imposed by the CAA Amendments on transportation planning came from alterations in the provisions on "transportation-air quality conformity." Under the CAA of 1977, conformity generally prohibited direct inconsistencies between long-range transportation system plans and the state implementation plans developed by states to come into compliance with the NAAQS. The CAA Amendments changed the conformity provisions by requiring that transportation plans conform to "a SIP's purpose of eliminating and reducing the severity and number of violations of the NAAQS and achieving expeditious attainment of the standards" (42 U.S.C. sec. 7506(c)(1)(A)). Hence, instead of merely avoiding direct conflicts between transportation and air quality plans, the new conformity provisions created an affirmative duty for transportation planners to create plans that would work to achieve the aims of providing citizens with healthy air.

Although not specifically targeted at the tension between mobility and accessibility, the CAA conformity provisions and their implementing regulations gave at least some impetuous for engaging the issues. The most direct, and obvious, point for this engagement was the selection of transportation projects to be included in long-range system plans. A 1999 EPA-sponsored study, however, shows that in only a handful of high-growth metropolitan areas has the conformity process led to the scaling back or elimination of proposed highway projects and the promotion of transit investments (Howitt & Moore, 1999). Aside from Atlanta's high-profile conformity lapse settlement agreement (McCarthy, 2004), conformity's impact on project selection has been difficult to detect. A later study by the Congressional Research Service shows that whereas 63 nonattainment or maintenance areas experienced conformity lapses between 1997 and 2004, "[m]ost of these areas . . . returned to conformity quickly without major effects on their transportation programs: . . . only 5 areas had to change transportation plans in order to resolve a conformity lapse" (McCarthy, 2004, p. i).

As with project selection, conformity's influence on land use policy, while notable, has not been widespread. Although the Clean Air Act specifically disavows any "infringement on the existing authority of counties and cities to plan or control land use," many had hoped that the CAA's restrictive conformity requirements would lead to "tighter coordination of land use and transportation planning to promote development patterns that require less travel" (Howitt & Moore, 1999, p. 80). The EPA has finessed the tension between statutory prohibition and popular expectation by sponsoring research and providing specific guidance on how land use policies might fit into air quality planning and conformity analyses (Apogee Research, Inc. & Hagler Bailly, 1998; Jack Faucett Associates, 1999; Johnston, Rodier, Choy, & Abraham, 2000; U.S. EPA, 2001), but the agency stresses that its efforts are advisory only. Government agencies in several metropolitan areas have, in fact, used EPA's research and guidance to take air-quality credit for land use initiatives; leading examples include the Atlantic Steel redevelopment project in Atlanta, Portland, Oregon's metropolitan growth management policies, and the Chicago region's reassessment of infill development potential. Probably the most dramatic change credited to the influence of the conformity process, at least on land use institutions if not directly on land use policies, was the creation in 1999 of the Georgia Regional Transportation

Authority (GRTA), which, at least on paper, has substantial veto power over land use and transportation decisions in the Atlanta region (Nelson, 2000; Bullard, Johnson, & Torres, 2001), although it has been reluctant to fully exercise that authority (Filipova, 2006).

In addition to voluntary actions, the conformity implementing regulations contain mandatory regional emissions analysis standards that, indirectly, influence land use policies. In nonattainment and maintenance areas classified as serious or worse for ozone or carbon monoxide and that are greater than 200,000 in population, agencies preparing long-range transportation system plans must ensure that future land use development assumptions are consistent with the transportation system alternatives under consideration, effectively codifying a key holding from a conformity-based lawsuit brought against the Bay Area's Metropolitan Transportation Commission in the early 1990s (Garrett & Wachs, 1996). In other words, planners must account for the induced development effects that might be associated with different investment decisions (Johnston, 2004). This, in essence, requires the consideration of integrated, internally consistent land use–transportation scenarios. Consideration of changes in land use policies to increase accessibility, however, is not required.

ISTEA, TEA-21, SAFETEA-LU, MAP-21, FAST Act

The Intermodal Surface Transportation Efficiency Act of 1991 (ISTEA), adopted shortly after the CAA Amendments, promised a more fundamental realignment in transportation systems planning processes. Although actual on-the-ground impacts of ISTEA were slight, the Act at least facilitated a framework for shifting to an accessibility-based planning system.

ISTEA changed system planning processes through two primary mechanisms. First, it changed the rules about funding eligibility for different transportation modes. Prior transportation authorization bills contained strict modally based funding categories that allowed minimal crossover from one category/mode to the next (Katz, Puentes, & Bernstein, 2003). For example, ISTEA's predecessor, the Surface Transportation and Uniform Relocation Assistance Act of 1987, contained more than 14 different highway construction funding categories, four bridge construction programs, two highway safety programs, and four transit programs. Each of these funding programs contained strict definitions of the types of transportation projects that could be funded through the program, allowing little or no flexibility to use the funds for other types of projects; for the most part, variations from the strict allocation rules required specific legislative authorization. This high level of inflexibility created problems for local and metropolitan transportation planners. Not infrequently, the greatest local/regional transportation needs did not match up with the categories and funding priorities in the federal bill. With the overwhelming amount of total funding allocated to highway construction categories, regions were essentially induced to develop highway expansion solutions to transportation problems, even when other approaches may have been more appropriate and effective.

ISTEA marked a significant departure from this practice by substantially increasing funding flexibility: the three largest funding programs in the Act – the National Highway System, the Surface Transportation Program, and §9 transit formula funds – totaling nearly $60 billion (nearly 40% of funding in the entire bill) could be used to fund virtually any surface transportation capital improvement, regardless of mode. This step toward mode neutrality in federal funding was intended to at least partially "unload the dice" in local and regional transportation planning: Federal funding restrictions would no longer stand in the way of the most appropriate solution to transportation problems:

> The Committee feels that one of the most important things this legislation can do is give state and local officials the flexibility to make the crucial decisions on how their funds should be used. They will have the ability to choose the best transportation solution without the artificial constraints of funding categories.
>
> (House Committee on Transportation and Infrastructure, 1991, p. 1531)

An additional method introduced in ISTEA to promote mode neutrality was to equalize federal match ratios across modes at 80 percent. Prior to ISTEA, highway construction projects enjoyed an 80–90 percent federal match ratio, whereas transit was limited to 75 percent and on many projects was considerably lower. Whereas the increased flexibility authorized substantial reallocations of funds away from roads to transit and other alternative modes, states and MPOs actually shifted very few dollars to nonhighway projects – the overwhelming majority of the money stayed on the highway side of the ledger – and, despite Congress' intention of flattening out inequities in federal match ratios across modes, the imbalances have continued pretty much the same as they were (Surface Transportation Policy Project, 2000).

The second major change made by ISTEA was the significant expansion of factors that had to be considered in long-range transportation planning processes for metropolitan areas. This expansion required both the incorporation of a broader range of issues and more in-depth analysis. Prior to ISTEA, federal legislation required only that metropolitan transportation planning be continuous, cooperative, and coordinated. ISTEA required consideration of no fewer than 15 detailed factors, covering a number of diverse topics, including facility preservation, international border crossings, and facility security. Of particular importance was Factor 4: "The likely effect of transportation policy decisions on land use and development and the consistency of transportation plans and programs with the provisions of all applicable short- and long-term land use and development plans." According to the

Senate Committee on Environment and Public Works (1991, p. 29), which developed this language, the intent was for MPOs to

> demonstrate that capacity expansion will not be accompanied by increased development in a manner that will frustrate the goals of expansion. MPO plans should also specify how land use plans may encourage any necessary travel demand reduction or encourage the use and financial viability of mass transportation and non-motorized travel.

As illuminated by the Committee explanation, the requirement of Factor 4 went much further than merely requiring recognition that land use patterns influence travel behaviors, as the prior 3C planning guidelines did. It called on MPOs to guard against planning and building facilities that might induce land development that could create excessive travel on the proposed facility. It also sought to have MPOs investigate future land use patterns that might reduce the need to travel. In other words, the objective was to have MPOs incorporate into their transportation system planning processes ways to increase accessibility while reducing the need for mobility – to increase exchange, while minimizing travel.

The Federal Highway Administration regulations implementing Factor 4, however, did not go as far as the Senate Committee explanatory language suggested, opting instead to focus on the need for transportation system plans to be consistent with local land use plans (FHWA, 1993). Mere consistency, of course, does not ensure incorporation of accessibility-related goals. If local land use plans call for increasingly dispersed and isolated development, all consistency would seem to require is that transportation plans contain enough highway lane miles to service the mobility needs generated by that development. In short, the regulations were more focused on process issues related to horizontal policy consistency than on the substance of facilitating accessibility.

ISTEA was followed in 1998 by the Transportation Equity Act for the 21st Century (TEA-21). Although the metropolitan planning provisions of TEA-21 followed the general pattern of ISTEA, the factors to be considered in the development of long-range plans were "streamlined" from 15 to seven. Gone from the list was consideration of land use patterns in any form. The only provision that even hinted at land use was Factor D, which implored MPOs to consider improved quality of life in their planning processes. The Conference Committee for TEA-21 explained that this general reference was intended to include "the interaction between transportation decisions and local land use decisions appropriate to each area," but the Committee hastened to add, MPOs are only encouraged to consider such interactions (House Committee on Transportation and Infrastructure, 1998, p. 113). This permissive intention was further communicated by the very next subsection of the Act, which explicitly barred court action to enforce consideration of any of the planning factors. Interestingly, the Federal Highway

Administration chose to not alter its planning regulations in response to TEA-21 and retained the pre-TEA-21 requirement urging mere consistency with local land use plans.

The legislation that replaced TEA-21 was hard-won and long overdue. Although the funding authorization provisions of TEA-21 expired in federal fiscal year 2003, it was not until August of 2005 before Congress finally adopted replacement legislation. As with TEA-21, the Safe, Accountable, Flexible, Efficient Transportation Equity Act: A Legacy for Users (SAFETEA-LU) was intended to continue many of the innovations created in ISTEA. SAFETEA-LU expanded slightly on the planning factors in TEA-21, adding one and providing slight modifications to the pre-existing seven factors. TEA-21's Factor D, now renumbered Factor E, was expanded to explicitly incorporate land use considerations: Long-range planning processes now need to consider projects and strategies that will "protect and enhance the environment, promote energy conservation, improve the quality of life, and promote consistency between transportation improvements and State and local planned growth and economic development patterns" (SAFETEA-LU, 2005, p. 1552). This new language more or less codified the Federal Highway Administration interpretation of the previous version of this factor from TEA-21. It did not, however, include ISTEA's requirement for considering possible land development impacts of planned transportation facilities. Nor did it include the Senate Committee on Environment and Public Works' understanding of the parallel factor from ISTEA – that planners should investigate ways to use land use planning and development to reduce travel or to encourage increased transit ridership. Although planners were required to consider projects and strategies that "increase the accessibility and mobility options available to people and for freight" (SAFETEA-LU, 2005, p. 1552) – that mandate was generally interpreted in ways that treat accessibility as synonymous with mobility (Volpe National Transportation Systems Center 2005).

The 2012 replacement of SAFETEA-LU – Moving Ahead for Progress in the 21st Century (MAP-21) – made no changes in the planning factors. It did, however, provide provisions that allow for – but not require – metropolitan-scale scenario planning processes (23 U.S.C. sec. 134(i)(4)), a topic which is addressed in Chapter 4. In addition, MAP-21 introduced a performance-based approach that seeks to link transportation investments to achieving goal-based targets in key areas ranging from infrastructure condition to freight movement (23 U.S.C. secs. 150(b), 134(h)(2)). As these important improvements in transportation planning policy make their way into federal agency regulations (NARA, 2014), they may provide the stage for the development of accessibility-oriented planning processes, and there is some evidence that these changes are starting to occur, at least within larger metropolitan areas (Proffitt, Bartholomew, Ewing, & Miller, 2015). Whether these innovations, which have been continued in the more recent Fixing America's Surface Transportation Act (FAST Act), will result in real change in planning processes on the ground, or whether their promise is greater than their reality – the fate of many other ISTEA- and post-ISTEA-related transportation policy reforms (Katz et al., 2003) – remains to be seen.

Conclusion

Cities are places for exchange. They provide the stage for human interactions of all kinds, from the mundane to the profound. These exchanges create the basis for meaning and form the foundation of culture. Given the centrality of these functions to human society, seeking systems that support and nurture exchanges by facilitating their occurrence with minimal cost and effort would seem paramount.

Yet, regional transportation planning has historically had an opposite objective. By myopically focusing on mobility, transportation planning efforts have fostered dispersion of exchange locations, increasing the amount of travel needed to access them. This has frequently resulted in overall decreases in accessibility to places of exchange, particularly for those in our society without ready access to an automobile. If transportation systems and the communities that are both served and affected by those systems are to be fully integrated in a planning process, the focus must shift from mobility to accessibility.

Photo: Keith Bartholomew.

3 Quantitative analysis of regional transportation plans

Introduction

To gauge where transportation practice currently stands on the accessibility versus mobility continuum – and a series of related questions – we conducted a study of regional transportation plans (RTPs) prepared by metropolitan planning organizations (MPOs) across the United States. The analysis took the form of both quantitative and qualitative assessments with the aim of capturing as accurate a picture of the state of metropolitan transportation planning practice as possible. The results of our qualitative analysis are reviewed in Chapter 6. In this chapter, we consider the quantitative information derived from more than 100 plans.

These plans constitute a mixed sample. Basically, we started with a list of the 90 largest urbanized areas in the nation, went to MPO websites, and retrieved RTPs where posted. Where RTPs were not posted, we dropped MPOs from the sample. To also include the far more numerous smaller urbanized areas (metropolitan areas smaller than 350,000 people), we pulled a random sample of small urbanized areas and repeated the process described above. A total of 105 RTPs are included in this mixed sample.

Once we had selected which plans to study, we needed to peruse the plans for quantitative data. We extracted data and placed them in a spreadsheet. The second step was to analyze the data. The data remained spotty at best, which precluded a statistical analysis relating travel outcomes to dollar inputs, our original intended purpose. Instead, we computed descriptive statistics and conducted some simple statistical tests, which suggest pitfalls and opportunities for MPOs.

Population and employment

Population and employment growth rates vary from urbanized area to urbanized area. Three urbanized areas are actually projected to lose population. Several are growing at rates of more than 2.5 percent per year. Time horizons also vary from place to place (see Table 3.1). Whereas the minimum time horizon, by law, is 20 years, many RTPs plan for transportation investments for closer to 30 years.

Table 3.1 Base and Horizon Years from RTPs

Urbanized Area	*Base Year*	*Horizon Year*
Akron, OH	2013	2035
Albany, NY	2011	2035
Albuquerque, NM	2015	2040
Allentown, PA	2011	2030
Atlanta, GA	2010	2030
Austin, TX	2010	2035
Bakersfield-Kem, CA	2014	2040
Baltimore, MD	2011	2035
Baton Rouge, LA	2013	2037
Berks County, PA	2010	2030
Berkshire County, MA	2013	2040
Birmingham, AL	2014	2040
Bismarck, ND	2010	2035
Boise, ID	2014	2040
Boston, MA	2015	2040
Bridgeport-Stamford, CT	2011	2040
Buffalo, NY	2014	2040
Burlington, VT	2013	2035
Cape Coral, FL	2010	2035
Charleston, SC	2010	2035
Charlotte, NC	2014	2035
Chattanooga, TN	2013	2035
Chicago, IL	2010	2040
Cincinnati, OH	2012	2040
Cleveland, OH	2012	2030
Colorado Springs, CO	2012	2035
Columbia, SC	2008	2035
Columbus, OH	2012	2035
Dallas-Ft. Worth, TX	2014	2035
Dayton, OH	2012	2040
Denver, CO	2011	2035
Des Moines, IA	2014	2050
Dover, DE	2013	2040
Durham, NC	2013	2040
El Paso, TX	2013	2040
Fargo, ND	2014	2035
Flagstaff, AZ	2009	2030
Fresno, CA	2014	2040
Ft. Lauderdale, FL	2014	2040

Urbanized Area	*Base Year*	*Horizon Year*
Grand Rapids, MI	2011	2035
Greenville, SC	2013	2035
Harrisburg, PA	2014	2040
Hartford, CT	2011	2035
Honolulu, HI	2011	2035
Houston, TX	2015	2035
Indianapolis, IN	2011	2035
Jacksonville, FL	2014	2040
Kansas City, MO	2015	2040
Knoxville, TN	2013	2040
Lancaster, PA	2012	2040
Lansing, MI	2015	2040
Las Vegas, NV	2012	2035
Little Rock, AR	2010	2030
Los Angeles, CA	2011	2035
Louisville-Jefferson County, KY	2014	2035
Madison, WI	2012	2035
McAllen, TX	2010	2035
Miami, FL	2014	2035
Milwaukee, WI	2006	2035
Minneapolis, MN	2014	2040
Nashville, TN	2010	2035
Newark, NJ	2013	2035
New Haven, CT	2015	2040
New York, NY	2015	2035
Oklahoma City, OK	2012	2030
Omaha, NE	2015	2040
Orlando, FL	2014	2040
Palm Beach, FL	2014	2040
Pensacola, FL	2010	2035
Philadelphia, PA	2013	2035
Phoenix, AZ	2014	2035
Pittsburgh, PA	2011	2035
Port St. Lucie, FL	2011	2035
Portland, ME	2011	2035
Portland, OR	2014	2035
Providence, RI	2012	2035
Provo-Orem, UT	2011	2040

(*Continued*)

Table 3.1 (Continued)

Urbanized Area	*Base Year*	*Horizon Year*
Raleigh, NC	2013	2040
Reno, NV	2013	2035
Richmond, VA	2014	2035
Rochester, NY	2011	2035
Sacramento, CA	2013	2035
St. Louis, MO	2011	2035
Salt Lake City-Ogden, UT	2015	2040
San Antonio, TX	2014	2040
San Diego, CA	2011	2030
San Francisco-Oakland, CA	2013	2040
Santa Barbara, CA	2013	2030
Sarasota-Bradenton, FL	2014	2035
Savannah, GA	2014	2035
Scranton, PA	2011	2035
Seattle, WA	2010	2040
Spokane, WA	2013	2040
Springfield, MA	2012	2035
Stockton, CA	2014	2040
Tallahassee, FL	2010	2035
Tampa-St. Petersburg, FL	2014	2040
Toledo, OH	2011	2035
Tucson, AZ	2012	2040
Tulsa, OK	2012	2035
Virginia Beach, VA	2012	2034
Washington, DC	2014	2040
Wichita, KS	2010	2035
Winston-Salem, NC	2013	2035
Worcester, MA	2012	2035
Youngstown, OH	2013	2040

The net effect is substantial variation in population and employment growth during the period of the RTPs (see Tables 3.2 and 3.3 and Figures 3.1 and 3.2). At one extreme is the Youngstown urbanized area, projected to lose 7 percent of its population over a 27-year period. At the other extreme is the Austin urbanized area, projected to increase in population by 123 percent over 25 years.

For urbanized areas in our sample, the mean population growth during the planning period is 35.0 percent, and the median population growth is

Table 3.2 Population of Urbanized Areas (from RTPs)

Urbanized Area	*Population Base*	*Population Horizon*	*Population Growth (%)*
Akron, OH	710,357	730,247	2.8
Albany, NY	837,967	888,312	6
Albuquerque, NM	767,647	1,331,139	73.4
Allentown, PA	644,348	767,856	19.2
Atlanta, GA	5,260,436	7,895,193	50.1
Austin, TX	1,458,600	3,250,600	122.9
Bakersfield, CA	856,000	1,444,100	68.7
Baltimore, MD	2,662,691	2,958,500	11.1
Baton Rouge, LA	661,142	891,030	34.8
Berks County, PA	397,537	446,582	12.3
Berkshire County, MA	131,219	130,251	-0.7
Birmingham, AL	1,105,132	1,305,000	18.1
Bismarck, ND	94,719	130,577	37.9
Boise, ID	581,288	1,022,000	75.8
Boston, MA	3,163,900	3,601,600	13.8
Bridgeport-Stamford, CT	312,620	329,521	5.4
Buffalo, NY	1,135,509	1,328,730	17
Burlington, VT	156,545	205,445	31.2
Cape Coral, FL	523,866	1,034,400	97.5
Charleston, SC	647,194	835,534	29.1
Charlotte, NC	1,248,900	2,026,700	62.3
Chattanooga, TN	442,628	558,063	26.1
Chicago IL	8,608,208	11,000,000	27.8
Cincinnati, OH	1,945,882	2,385,497	22.6
Cleveland, TN	98,520	131,212	33.2
Colorado Springs, CO	645,429	955,191	48
Columbia, SC	576,101	845,664	46.8
Columbus, OH	1,400,000	1,900,000	35.7
Dallas-Ft. Worth, TX	6,778,201	9,833,378	45.1
Dayton, OH	799,732	820,227	2.6
Denver, CO	2,885,100	4,348,700	50.7
Des Moines, IA	480,000	751,000	56.5
Dover, DE	151,234	183,576	21.4
Durham, NC	1,620,035	2,929,976	80.9
El Paso, TX	832,836	1,158,195	39.1
Fargo, ND	208,777	298,070	42.8
Flagstaff, AZ	79,383	103,621	30.5

(*Continued*)

Table 3.2 (Continued)

Urbanized Area	*Population Base*	*Population Horizon*	*Population Growth (%)*
Fresno, CA	912,521	1,373,679	50.5
Ft. Lauderdale, FL	1,748,800	1,962,000	12.2
Grand Rapids, MI	648,139	923,727	42.5
Greenville, SC	621,834	840,861	35.2
Harrisburg, PA	549,475	627,469	14.2
Hartford, CT	769,598	826,000	7.3
Honolulu, HI	905,600	1,136,000	25.4
Houston, TX	5,900,000	9,200,000	55.9
Indianapolis, IN	1,834,935	2,365,551	28.9
Jacksonville, FL	1,262,795	1,964,600	55.6
Kansas City, MO	1,895,551	2,492,030	31.5
Knoxville, TN	893,507	1,300,000	45.5
Lancaster, PA	519,445	652,000	25.5
Lansing, MI	454,667	491,808	8.2
Las Vegas, NV	1,951,269	2,950,000	51.2
Little Rock, AR	1,118,536	1,577,296	41
Los Angeles, CA	18,616,290	22,140,614	18.9
Louisville-Jefferson County, KY	947,150	1,151,805	21.6
Madison, WI	488,073	653,900	34
McAllen, TX	628,000	1,072,000	70.7
Miami, FL	2,500,000	3,300,000	32
Milwaukee, WI	1,391,200	2,276,000	63.6
Minneapolis, MN	2,850,000	3,673,860	28.9
Nashville, TN	1,394,928	2,174,914	55.9
Newark, NJ	6,578,900	7,910,400	20.2
New York, NY	12,600,000	14,300,000	13.5
Oklahoma City, OK	1,076,258	1,464,814	36.1
Omaha, NE	742,146	947,500	27.7
Orlando, FL	2,604,527	2,877,916	10.5
Palm Beach, FL	1,320,000	1,680,000	27.3
Pensacola, FL	911,896	1,153,800	26.5
Philadelphia, PA	5,626,186	6,261,673	11.3
Phoenix, AZ	4,062,543	6,258,452	54.1
Pittsburgh, PA	2,574,959	3,056,715	18.7
Port St. Lucie, FL	377,055	648,600	72
Portland, ME	312,866	393,881	25.9
Portland, OR	1,483,506	2,080,456	40.2
Providence, RI	1,048,319	1,140,543	8.8

Urbanized Area	*Population Base*	*Population Horizon*	*Population Growth (%)*
Provo-Orem, UT	516,564	1,092,450	111.5
Raleigh, NC	1,060,192	1,989,641	87.7
Reno, NV	421,407	573,114	36
Richmond, VA	1,001,650	1,437,704	43.5
Rochester, NY	1,225,054	1,261,141	3
Sacramento, CA	2,215,044	3,086,213	39.3
St. Louis, MO	2,571,253	2,804,244	9.1
Salt Lake City-Ogden, UT	1,512,017	2,401,940	58.9
San Antonio, TX	1,988,188	3,404,143	71.2
San Diego, CA	3,131,552	4,384,867	40
San Francisco-Oakland, CA	7,150,740	9,299,150	30
Santa Barbara, CA	423,895	519,965	22.7
Sarasota-Bradenton, FL	723,900	999,643	38.1
Savannah, GA	276,434	320,000	15.8
Scranton, PA	532,545	564,497	6
Seattle, WA	3,675,000	4,970,000	35.2
Spokane, WA	471,221	636,000	35
Springfield, MA	621,570	658,000	5.9
Stockton, CA	704,794	1,070,486	51.9
Tallahassee, FL	47,700	58,221	22.1
Tampa-St. Petersburg, FL	1,229,266	1,815,964	47.7
Toledo, OH	627,064	618,812	-1.3
Tucson, AZ	1,000,000	1,800,000	80
Tulsa, OK	778,051	1,030,735	32.5
Virginia Beach, VA	1,600,000	2,000,000	25
Washington, DC	7,057,312	8,794,590	24.6
Wichita, KS	112,700	124,212	10.2
Winston-Salem, NC	393,018	552,867	40.7
Worcester, MA	556,698	640,000	15
Youngstown, OH	449,135	416,601	-7.2

31.5 percent. The mean employment growth is 37.7 percent, and the median is 36.4 percent. The significance of these fairly large numbers is that in the typical urbanized area, more than a quarter of the development on the ground in the horizon year will have been built during the planning period. Possibly more, with redevelopment. This presents an opportunity. The new development can be compact and multimodal, or it can be sprawling and auto-dependent.

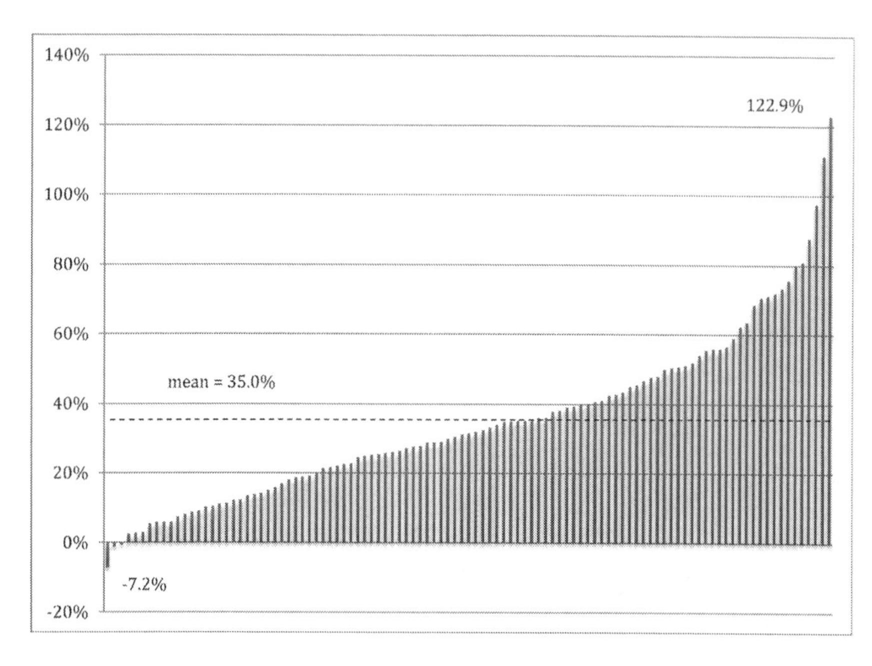

Figure 3.1 Population Growth during Forecast Period.

Table 3.3 Employment of Urbanized Areas (from RTPs)

Urbanized Area	*Employment Base*	*Employment Horizon*	*Employment Growth (%)*
Akron, OH	299,868	344,301	14.8
Albany, NY	450,000	500,000	11.1
Albuquerque, NM	396,477	586,205	47.9
Allentown, PA	644,348	767,856	19.2
Atlanta, GA	3,098,828	4,715,620	52.2
Austin, TX	698,400	1,642,800	135.2
Bakersfield, CA	325,000	500,000	53.8
Baltimore, MD	1,686,600	1,948,800	15.5
Baton Rouge, LA	330,560	441,740	33.6
Berks County, PA	181,000	235,000	29.8
Berkshire County, MA	60,150	55,650	-7.5
Birmingham, AL	350,279	421,960	20.5
Bismarck, ND	52,100	73,300	40.7
Boise, ID	240,506	461,657	92
Boston, MA	1,850,000	2,027,800	9.6
Bridgeport-Stamford, CT	136,399	165,845	21.6
Buffalo, NY	638,758	714,050	11.8

Urbanized Area	*Employment Base*	*Employment Horizon*	*Employment Growth (%)*
Burlington, VT	122,458	171,783	40.3
Cape Coral, FL	278,203	440,334	58.3
Charleston, SC	303,488	392,736	29.4
Charlotte, NC	829,800	1,304,900	57.3
Chattanooga, TN	216,001	299,680	38.7
Cincinnati, OH	974,804	1,223,135	25.5
Cleveland, TN	48,810	68,237	39.8
Colorado Springs, CO	350,000	520,000	48.6
Columbia, SC	304,974	446,821	46.5
Columbus, OH	750,000	950,000	26.7
Dallas-Ft. Worth, TX	4,292,516	6,177,016	43.9
Dayton, OH	441,393	463,633	5
Denver, CO	1,561,400	2,576,000	65
Des Moines, IA	338,000	481,000	42.3
Dover, DE	73,802	80,956	9.7
Durham, NC	532,365	841,164	58
El Paso, TX	306,656	429,455	40
Fargo, ND	118,266	162,429	37.3
Flagstaff, AZ	36,815	53,969	46.6
Fresno, CA	345,816	449,111	29.9
Ft. Lauderdale, FL	959,800	1,059,619	10.4
Grand Rapids, MI	427,896	501,896	17.3
Harrisburg, PA	313,347	378,560	20.8
Honolulu, HI	556,900	693,300	24.5
Houston, TX	2,700,000	4,200,000	55.6
Indianapolis, IN	868,129	1,034,395	19.2
Jacksonville, FL	642,365	985,007	53.3
Kansas City, MO	938,052	1,260,571	34.4
Knoxville, TN	576,987	888,498	54
Lansing, MI	280,786	299,647	6.7
Las Vegas, NV	810,000	1,200,000	48.1
Little Rock, AR	292,212	420,561	43.9
Los Angeles, CA	7,500,000	9,500,000	26.7
Louisville-Jefferson County, KY	536,705	868,648	61.8
Madison, WI	255,800	364,400	42.5
McAllen, TX	158,000	292,000	84.8
Miami, FL	1,375,321	1,994,215	45
Milwaukee, WI	1,179,000	1,186,900	0.7
Minneapolis, MN	1,600,000	2,150,000	34.4

(*Continued*)

Table 3.3 (Continued)

Urbanized Area	*Employment Base*	*Employment Horizon*	*Employment Growth (%)*
Nashville, TN	983,074	1,536,746	56.3
Newark, NJ	2,847,400	3,771,700	32.5
New York, NY	7,200,000	8,900,000	23.6
Oklahoma City, OK	578,306	801,302	38.6
Omaha, NE	440,000	560,000	27.3
Orlando, FL	1,126,741	1,678,985	49
Palm Beach, FL	571,000	820,000	43.6
Pensacola, FL	213,132	253,223	18.8
Philadelphia, PA	2,950,387	3,268,881	10.8
Phoenix, AZ	1,736,465	3,023,732	74.1
Pittsburgh, PA	1,600,000	1,950,000	21.9
Portland, ME	294,819	350,828	19
Portland, OR	755,337	1,185,794	57
Port St. Lucie, FL	150,319	246,796	64.2
Providence, RI	467,299	516,048	10.4
Provo-Orem, UT	283,915	560,058	97.3
Raleigh, NC	260,000	430,000	65.4
Reno, NV	232,681	360,932	55.1
Richmond, VA	519,660	722,943	39.1
Rochester, NY	530,000	545,900	3
Sacramento, CA	966,316	1,327,424	37.4
Salt Lake City-Ogden, UT	915,297	1,373,996	50.1
San Antonio, TX	870,476	1,674,457	92.4
San Diego, CA	1,501,080	2,003,038	33.4
San Francisco-Oakland, CA	3,385,800	4,505,220	33.1
Santa Barbara, CA	197,400	257,600	30.5
Sarasota-Bradenton, FL	333,895	465,195	39.3
Savannah, GA	107,783	141,279	31.1
Seattle, WA	1,891,000	2,710,000	43.3
Spokane, WA	259,999	267,000	2.7
Springfield, MA	194,546	262,576	35
Stockton, CA	206,881	299,717	44.9
Tampa-St. Petersburg, FL	711,400	1,112,059	56.3
Tucson, AZ	375,000	740,000	97.3
Tulsa, OK	460,917	568,156	23.3
Virginia Beach, VA	1,000,875	1,200,000	19.9
Washington, DC	4,126,629	5,520,106	33.8
Wichita, KS	63,200	71,100	12.5
Winston-Salem, NC	244,079	333,040	36.4
Worcester, MA	224,000	250,000	11.6
Youngstown, OH	250,548	270,887	8.1

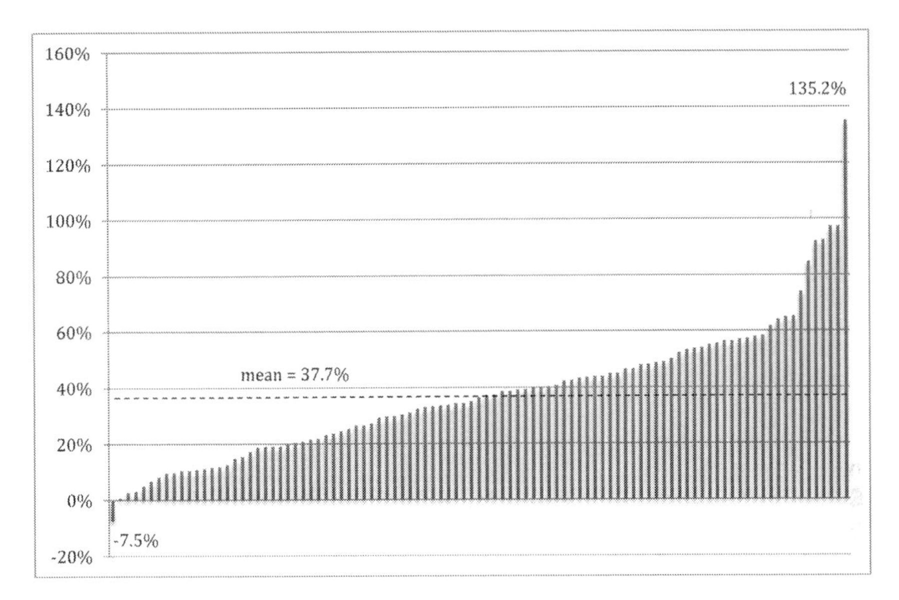

Figure 3.2 Employment Growth during Forecast Period.

VMT

For our sample of RTPs, total VMT is projected to grow in all urbanized areas, even in those few that are losing population (see Table 3.4 and Figure 3.3). The greatest increment of growth is 121 percent over 29 years in Provo-Orem, UT. The mean growth increment is 41 percent, the median is 35.6 percent.

Part of this VMT growth is due to population and employment growth itself. The simple correlation between the population growth rate and the VMT growth rate is 0.60. However, in 50 of 80 areas for which data are available, VMT is projected to grow faster than population, suggesting that other factors are at work. Two of these other factors may be urban sprawl and highway expansion (see Chapters 4 and 5).

We divided total VMT by total population to obtain VMT per capita for different regions (see Table 3.5 and Figure 3.4). The VMT per capita estimates range so widely from place to place we suspect that VMT is estimated on different bases in different places. Still, we would expect that any given MPO uses a consistent methodology for the base and horizon years, and therefore that relative changes over time are representative.

Even on a per capita basis, VMT is projected to increase in most regions, and substantially in highway-oriented regions such as Orlando and Springfield. One region that projects a significant decrease in VMT per capita is Portland, OR, and even its decrease is not enough to meet county or state climate goals.

Table 3.4 Aggregate VMT of Urbanized Areas (from RTPs)

Urbanized Area	*VMT Base*	*VMT Horizon*	*VMT Growth (%)*
Akron, OH	20,000,000	25,000,000	25
Albany, NY	27,738,000	37,760,000	36.1
Albuquerque, NM	16,288,169	31,554,951	93.7
Allentown, PA	13,390,000	19,600,000	46.4
Atlanta, GA	17,377,388	22,146,969	27.4
Austin, TX	40,897,681	73,063,023	78.6
Bakersfield, CA	22,236,000	38,197,000	71.8
Baltimore, MD	61,508,400	82,069,700	33.4
Baton Rouge, LA	21,251,780	26,770,713	26
Birmingham, AL	27,694,974	44,113,162	59.3
Bismarck, ND	1,460,000	2,330,000	59.6
Boston, MA	69,448,500	74,970,100	8
Buffalo, NY	20,500,000	31,087,235	51.6
Burlington, VT	5,142,001	6,104,605	18.7
Cape Coral, FL	26,484,159	46,287,668	74.8
Charleston, SC	16,563,373	21,532,385	30
Charlotte, NC	38,566	46,000	19.3
Chattanooga, TN	11,154,226	15,235,120	36.6
Cincinnati, OH	56,608,044	76,440,176	35
Colorado Springs, CO	20,108,186	29,315,918	45.8
Columbus, OH	34,600,260	43,492,639	25.7
Dallas-Ft. Worth, TX	181,516,746	281,580,581	55.1
Dayton, OH	21,291,000	27,765,000	30.4
Denver, CO	74,435,100	119,255,207	60.2
Dover, DE	4,650,000	7,810,000	68
Durham, NC	13,217,550	20,884,276	58
El Paso, TX	15,787,118	20,559,661	30.2
Flagstaff, AZ	2,329,284	3,373,327	44.8
Fresno, CA	12,592,900	18,285,600	45.2
Grand Rapids, MI	16,478,493	20,009,105	21.4
Greenville, SC	12,063,573	16,364,490	35.7
Harrisburg, PA	15,168,442	21,113,219	39.2
Hartford, CT	21,000,000	27,490,000	30.9
Honolulu, HI	13,142,600	15,208,900	15.7
Houston, TX	160,000,000	275,000,000	71.9
Jacksonville, FL	43,145,248	67,105,192	55.5
Kansas City, MO	51,674,878	69,989,075	35.4
Knoxville, TN	25,153,956	40,302,079	60.2
Lansing, MI	3,789,000	4,081,000	7.7
Las Vegas, NV	39,978,598	54,620,983	36.6

Urbanized Area	*VMT Base*	*VMT Horizon*	*VMT Growth (%)*
Little Rock, AR	21,361,309	30,690,000	43.7
Los Angeles, CA	445,843,000	548,374,000	23
Louisville-Jefferson County, KY	33,905,000	43,913,000	29.5
Miami, FL	43,822,000	60,000,000	36.9
Milwaukee, WI	40,350,000	47,000,000	16.5
Minneapolis, MN	72,900,000	89,420,000	22.7
Newark, NJ	144,700,000	202,580,000	40
New Haven, CT	14,330,357	15,501,526	8.2
New York, NY	176,944,390	198,757,939	12.3
Oklahoma City, OK	34,000,000	47,500,000	39.7
Orlando, FL	60,273,972	122,617,840	103.4
Palm Beach, FL	11,047,137	14,893,927	34.8
Pensacola, FL	29,248,797	39,478,538	35
Philadelphia, PA	113,789,800	133,928,500	18.7
Phoenix, AZ	75,682,091	122,928,512	62.4
Pittsburgh, PA	62,783,674	77,445,008	23.4
Portland, ME	23,326,384	26,530,068	13.7
Portland, OR	19,226,604	25,307,208	31.6
Provo-Orem, UT	9,998,092	22,088,601	120.9
Raleigh, NC	31,018,970	56,644,594	82.6
Reno, NV	10,200,000	14,500,000	42.2
Richmond, VA	40,090,645	50,559,447	26.1
Rochester, NY	13,320,089	17,441,962	30.9
Sacramento, CA	57,230,800	74,308,300	29.8
St. Louis, MO	65,841,928	85,984,505	30.6
Salt Lake City-Ogden, UT	37,800,425	64,852,380	71.6
San Antonio, TX	48,896,300	98,486,800	101.4
San Diego, CA	85,072,545	121,522,502	42.8
Santa Barbara, CA	8,990,000	12,510,000	39.2
Sarasota-Bradenton, FL	19,206,847	25,746,825	34.1
Scranton, PA	15,437,429	17,992,017	16.5
Seattle, WA	78,461,538	102,000,000	30
Spokane, WA	8,408,329	11,306,096	34.5
Springfield, MA	16,000,000	28,000,000	75
Stockton, CA	18,113,206	29,065,851	60.5
Toledo, OH	14,179,919	14,949,877	5.4
Tucson, AZ	60,000,000	126,000,000	110
Tulsa, OK	29,093,000	36,604,000	25.8
Washington, DC	167,728,795	207,557,300	23.7
Winston-Salem, NC	26,450,630	37,892,388	43.3
Youngstown, OH	16,466,293	17,372,006	5.5

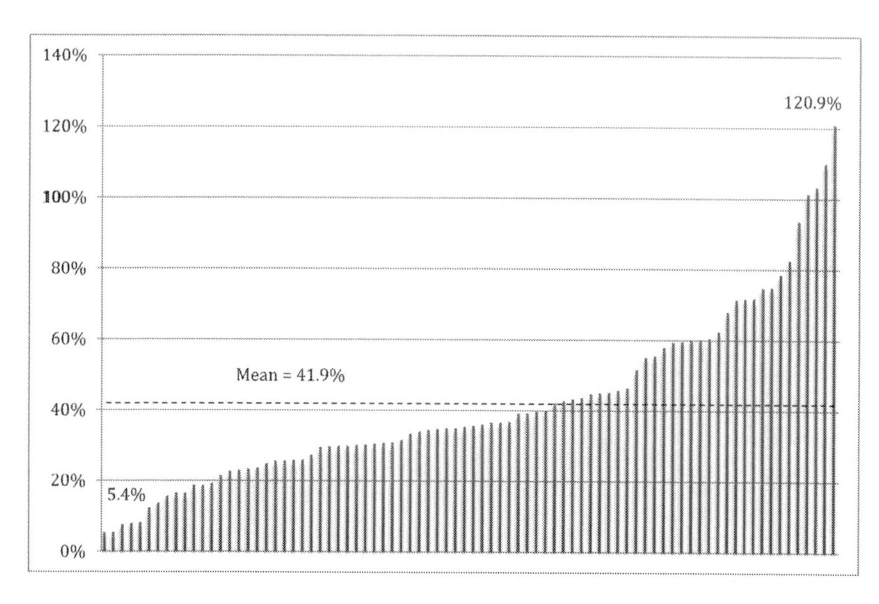

Figure 3.3 VMT Growth during Forecast Period.

Table 3.5 VMT per Capita of Urbanized Areas (from RTPs)

Urbanized Area	*VMT/Cap Base*	*VMT/Cap Horizon*	*VMT/Cap Growth (%)*
Akron, OH	28.15	34.23	21.6
Albany, NY	33.1	42.51	28.4
Albuquerque, NM	21.22	23.71	11.7
Allentown, PA	20.78	25.53	22.8
Austin, TX	28.04	22.48	-19.8
Bakersfield, CA	25.98	26.45	1.8
Baltimore, MD	23.1	27.74	20.1
Baton Rouge, LA	32.14	30.04	-6.5
Birmingham, AL	25.06	33.8	34.9
Bismarck, ND	15.41	17.84	15.8
Boston, MA	22	20.8	-5.5
Buffalo, NY	18.05	23.4	29.6
Burlington, VT	32.8	29.7	-9.4
Cape Coral, FL	50.56	44.75	-11.5
Charleston, SC	25.59	25.77	0.7
Charlotte, NC	30.9	22.7	-26.5
Chattanooga, TN	25.2	27.3	8.3

Urbanized Area	*VMT/Cap Base*	*VMT/Cap Horizon*	*VMT/Cap Growth (%)*
Cincinnati, OH	29.1	32	10.1
Colorado Springs, CO	31.15	30.69	-1.5
Columbus, OH	24.71	22.89	-7.4
Dallas-Ft. Worth, TX	26.78	28.64	6.9
Dayton, OH	26.62	33.85	27.2
Denver, CO	25.8	27.42	6.3
Dover, DE	30.75	42.54	38.4
Durham, NC	8.16	7.13	-12.6
El Paso, TX	18.96	17.75	-6.4
Flagstaff, AZ	29.3	32.6	11.3
Fresno, CA	13.8	13.31	-3.5
Grand Rapids, MI	25.42	21.66	-14.8
Greenville, SC	19.4	19.46	0.3
Harrisburg, PA	27.61	33.65	21.9
Hartford, CT	27.29	33.28	22
Honolulu, HI	14.51	13.39	-7.8
Houston, TX	27.12	29.89	10.2
Jacksonville, FL	34.17	34.16	0
Kansas City, MO	27.3	28.1	3
Knoxville, TN	28.15	31	10.1
Lansing, MI	8.33	8.3	-0.4
Las Vegas, NV	20.49	18.52	-9.6
Little Rock, AR	19.1	19.46	1.9
Los Angeles, CA	23.95	24.77	3.4
Louisville-Jefferson County, KY	35.8	38.13	6.5
Miami, FL	17.53	18.18	3.7
Milwaukee, WI	29	20.65	-28.8
Minneapolis, MN	25.6	24.3	-5.1
New York, NY	14.04	13.9	-1
Newark, NJ	21.99	25.61	16.4
Oklahoma City, OK	31.59	32.43	2.7
Orlando, FL	23.14	42.61	84.1
Palm Beach, FL	25	20	-20
Pensacola, FL	32.07	34.22	6.7
Philadelphia, PA	20.23	21.39	5.8
Phoenix, AZ	18.63	19.64	5.4
Pittsburgh, PA	24.38	25.34	3.9
Portland, ME	74.56	67.36	-9.7

(*Continued*)

Table 3.5 (Continued)

Urbanized Area	*VMT/Cap Base*	*VMT/Cap Horizon*	*VMT/Cap Growth (%)*
Portland, OR	13.01	12.2	-6.1
Providence, RI	23.1	24	3.8
Provo-Orem, UT	19.35	20.22	4.5
Raleigh, NC	29.26	28.47	-2.7
Reno, NV	24.2	25.3	4.5
Richmond, VA	40.02	35.17	-12.1
Rochester, NY	10.87	13.83	27.2
Sacramento, CA	25.84	24.08	-6.8
St. Louis, MO	25.61	30.66	19.7
Salt Lake City-Ogden, UT	25	27	8
San Antonio, TX	24.59	28.93	17.6
San Diego, CA	27.17	27.71	2
San Francisco-Oakland, CA	22	20	-9.1
Santa Barbara, CA	21.21	24.06	13.4
Sarasota-Bradenton, FL	26.53	25.76	-2.9
Scranton, PA	28.99	31.87	10
Seattle, WA	21.35	20.52	-3.9
Spokane, WA	17.84	17.78	-0.4
Springfield, MA	25.74	42.55	65.3
Stockton, CA	25.7	27.15	5.7
Toledo, OH	22.61	24.16	6.8
Tucson, AZ	60	70	16.7
Tulsa, OK	37.39	35.51	-5
Washington, DC	23.77	23.6	-0.7
Winston-Salem, NC	67.3	68.54	1.8
Youngstown, OH	36.66	41.7	13.7

There is some good news and some bad news. The projected rate of growth of VMT per capita is lower than its historical norm. Since 1980, VMT for the United States as a whole has grown almost three times faster than population. In contrast, for our sample, the mean percent projected increase in VMT per capita is only 5.8 percent over the period of the RTPs, whereas the median increase is 3.7 percent. Now for the bad news. When we conducted a paired-sample difference-of-means test, VMT per capita is still projected to increase to a statistically significant ($p = 0.002$) extent from base year to horizon year.

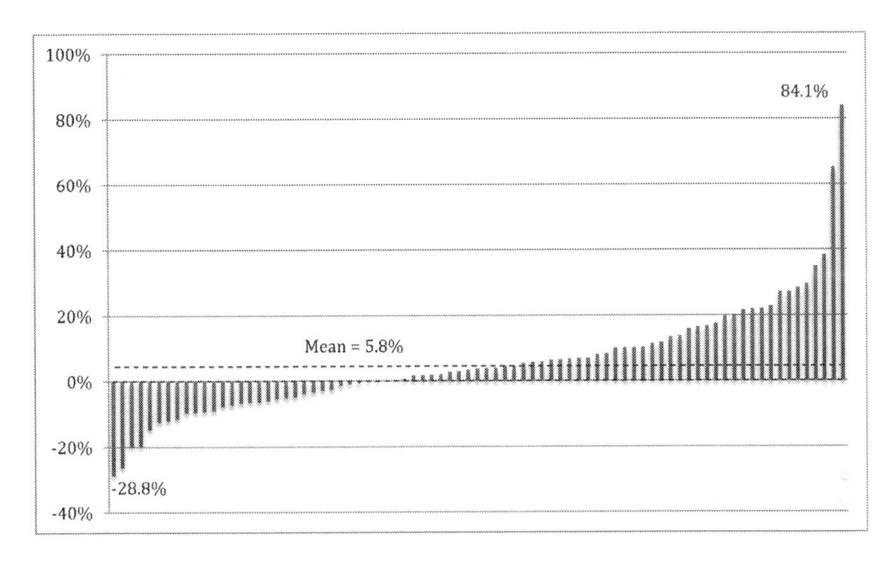

Figure 3.4 Change in VMT per Capita during Forecast Period.

Congestion

As the Raleigh and Durham RTPs state: "No region has been able to 'build its way' out of congestion; an important challenge for our transportation plans is to provide travel choices that allow people to avoid congestion where we cannot prevent it" (Capital Area Metropolitan Planning Organization and Durham-Chapel Hill-Carrboro Metropolitan Planning Organization, 2013, p. 3). This statement is echoed in the RTPs of others as well. Figure 3.5 illustrates the point.

For RTPs that include it, traffic delay is projected to grow in all but a few urbanized areas (see Table 3.6 and Figure 3.6). No matter how it is measured, congestion is projected to increase significantly between base and horizon years. This is often despite substantial increases in highway capacity.

For our sample, the mean increase in traffic delay is 237.7 percent over the forecast period, whereas the median is 139.6 percent. Delay is almost always projected to grow much faster than VMT. This is the case because delay rises exponentially as VMT approaches roadway capacity, and lane miles of capacity are not increasing as fast as VMT. Again, regions cannot pave their way out of congestion.

We computed traffic delay per capita, and as with VMT, it is clear from the wide range of values that delay is measured differently in different places (see Table 3.7 and Figure 3.7). Still, we can assume that any given

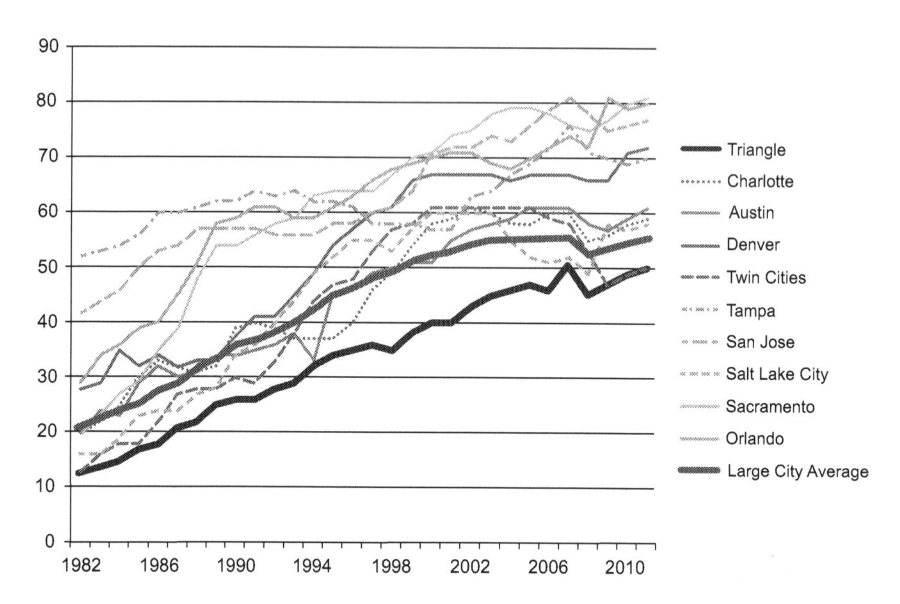

Figure 3.5 Percent of Peak Travel in Congestion.

(Capital Area Metropolitan Planning Organization and Durham-Chapel Hill-Carrboro Metropolitan Planning Organization, 2013, p. 17).

Table 3.6 Aggregate Delay for Urbanized Areas (from RTPs)

Urbanized Area	*Daily Hrs Delay Base*	*Daily Hrs Delay Horizon*	*Delay Growth*
Albany, NY	8,057	10,878	35.00%
Albuquerque, NM	8,919	137,618	1,443.00%
Atlanta, GA	101,722	323,544	218.10%
Austin, TX	151,394	836,748	452.70%
Bakersfield, CA	4,273	8,344	95.30%
Baltimore, MD	20,297	91,777	352.20%
Baton Rouge, LA	139,136	240,774	73.00%
Boise, ID	27,670	430,350	1,455.73%
Burlington, VT	3,097	6,146	98.50%
Cape Coral, FL	76,105	208,175	173.50%
Charleston, SC	45,983	91,965	100.00%
Chattanooga, TN	40,460	76,641	89.40%
Cincinnati, OH	40,529	189,980	368.80%
Cleveland, TN	13,100	22,900	74.80%
Columbus, OH	81,293	269,283	231.20%

Urbanized Area	*Daily Hrs Delay Base*	*Daily Hrs Delay Horizon*	*Delay Growth*
Dallas-Ft. Worth, TX	1,168,022	2,510,535	114.90%
Dayton, OH	5,046	16,861	234.10%
Denver, CO	200,430	628,290	213.50%
Durham, NC	27,446	139,455	408.10%
Fresno, CA	80,623	684,633	749.20%
Honolulu, HI	90,900	100,100	10.10%
Houston, TX	1,631,322	2,704,093	65.80%
Indianapolis, IN	35,000	87,000	148.60%
Knoxville, TN	97,654	299,378	206.60%
Los Angeles, CA	1,253	5,133	309.70%
Madison, WI	48,315	142,012	193.90%
Miami, FL	380,000	1,580,000	315.80%
New York, NY	4,400,000	6,100,000	38.60%
Oklahoma City, OK	57,107	114,697	100.80%
Pensacola, FL	29,820	74,833	150.90%
Philadelphia, PA	886,980	1,096,006	23.60%
Phoenix, AZ	735,398	1,715,043	133.20%
Pittsburgh, PA	69,084	168,607	144.10%
Port St. Lucie, FL	3,712	10,530	183.70%
Raleigh, NC	68,576	231,744	237.90%
Sacramento, CA	24,506	27,781	13.40%
St. Louis, MO	426,075	584,281	37.10%
Salt Lake City-Ogden, UT	62,132	140,647	126.40%
San Antonio, TX	1,056	11,940	1030.20%
San Diego, CA	156,578	363,944	132.40%
Santa Barbara, CA	15,627	24,534	57.00%
Sarasota-Bradenton, FL	192,891	462,219	139.60%
Seattle, WA	954,289	1,033,002	8.20%
Tallahassee, FL	29,000	51,000	75.90%
Tampa. FL	158,289	614,770	288.38%
Toledo, OH	16,234	16,702	2.90%

place uses a consistent methodology, and therefore that the relative growth of delay is accurate. Delay per capita increases by a mean value of 143.1 percent and a median value of 73.5 percent during the planning period. In a difference-of-means test, this increase is significant at the 0.001 probability level.

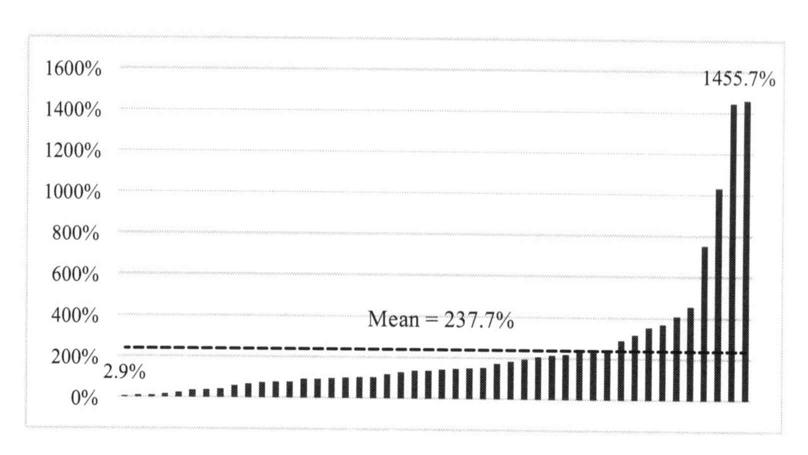

Figure 3.6 Delay Growth during Forecast Period.

Table 3.7 Delay per Capita for Urbanized Areas (from RTPs)

Urbanized Area	*Daily Delay per Capita Base*	*Daily Delay per Capita Horizon*	*Delay per Capita Growth*
Albany, NY	0.01	0.012	27.4
Albuquerque, NM	0.012	0.103	789.8
Atlanta, GA	0.019	0.041	111.9
Austin, TX	0.104	0.257	148
Bakersfield, CA	0.005	0.006	15.7
Baltimore, MD	0.008	0.031	307
Baton Rouge, LA	0.21	0.27	28.4
Boise, ID	0.048	0.431	806.5
Burlington, VT	0.02	0.03	50
Cape Coral, FL	0.145	0.201	38.5
Charleston, SC	0.071	0.11	54.9
Chattanooga, TN	0.091	0.137	50.2
Chicago IL	0.001	0.001	–60.8
Cincinnati, OH	0.021	0.08	282.4
Cleveland, TN	0.133	0.175	31.3
Columbus, OH	0.058	0.142	144.1
Dallas-Ft. Worth, TX	0.172	0.255	48.2
Dayton, OH	0.006	0.021	225.8
Denver, CO	0.069	0.144	108
Durham, NC	0.017	0.048	180.9
Fresno, CA	0.088	0.498	464.1
Honolulu, HI	0.1	0.088	–12.2
Houston, TX	0.276	0.294	6.3

Urbanized Area	*Daily Delay per Capita Base*	*Daily Delay per Capita Horizon*	*Delay per Capita Growth*
Indianapolis, IN	0.019	0.037	92.8
Knoxville, TN	0.109	0.23	110.7
Los Angeles, CA	0.255	0.19	–25.7
Madison, WI	0.099	0.217	119.4
Miami, FL	0.152	0.479	215
New York, NY	0.349	0.427	22.2
Oklahoma City, OK	0.053	0.078	47.6
Pensacola, FL	0.033	0.065	98.3
Philadelphia, PA	0.158	0.175	11
Phoenix, AZ	0.181	0.274	51.4
Pittsburgh, PA	0.027	0.055	105.6
Port St. Lucie, FL	0.01	0.016	64.9
Raleigh, NC	0.065	0.116	80.1
Sacramento, CA	0.011	0.009	–18.6
St. Louis, MO	0.166	0.208	25.7
Salt Lake City-Ogden, UT	0.041	0.059	42.5
San Antonio, TX	0.19	1.28	660
San Diego, CA	0.05	0.083	66
Santa Barbara, CA	0.037	0.047	28
Sarasota-Bradenton, FL	0.266	0.462	73.5
Seattle, WA	0.26	0.208	–20
Tallahassee, FL	0.608	0.876	44.1
Tampa, FL	0.13	0.34	161.5
Toledo, OH	0.026	0.027	4.3

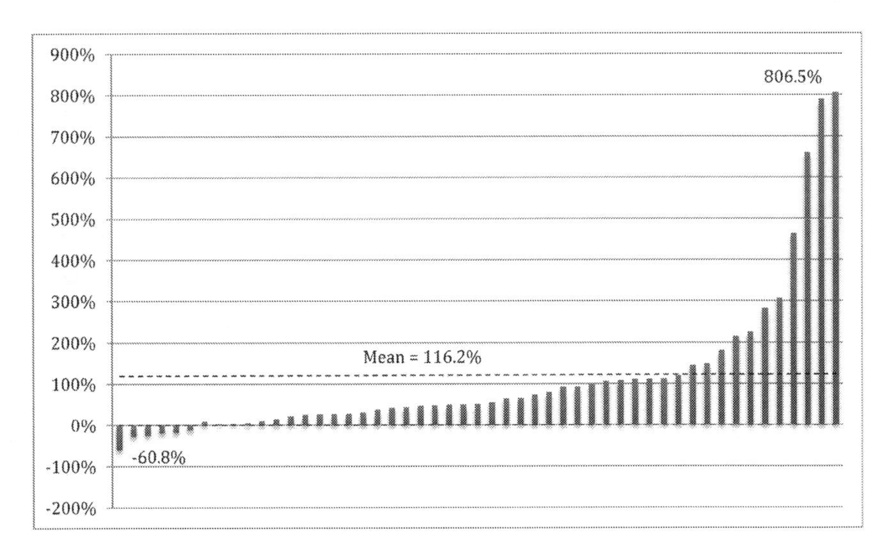

Figure 3.7 Change in Delay per Capita during Forecast Period.

Mode shares

In some of our regions, walking, biking, and transit use have been on an upswing (see Figure 3.8). One particularly bright spot in the performance picture is in the area of transit ridership. Transit mode share is projected to increase in all but six of the 28 regions for which projections are available (see Table 3.8 and Figure 3.9). However, there is great variation from place to place, reflecting expected land use patterns, changing demographics, and expected transit levels of service. Also reflected are different trip purposes within our sample. Transit has historically captured a larger share of trips for work than other purposes. Some of our mode share data are for work trips only. Others are for all trips.

Increases in transit mode share are dramatic in places such as Sacramento, CA, Provo-Orem, UT, and Palm Beach, FL. For our sample, the mean increase in transit mode share is 63.2 percent, the median is 17.6 percent. A difference-of-means test of proportions shows the increase in transit mode share to be statistically significant at the 0.002 probability level.

The prognosis is also favorable for walking and bicycling, although less so than for transit (see Table 3.9 and Figure 3.10). The walk/bike mode share is projected to increase in 18 of the 21 urbanized areas for which projections are available. The mean increase in walk mode share for our sample is 42.1 percent and the median increase is 11.2 percent. The mean increase is statistically significant at the 0.005 level.

There may be two reasons for the smaller increase in walking than transit use. One is that walking is even more sensitive to development patterns

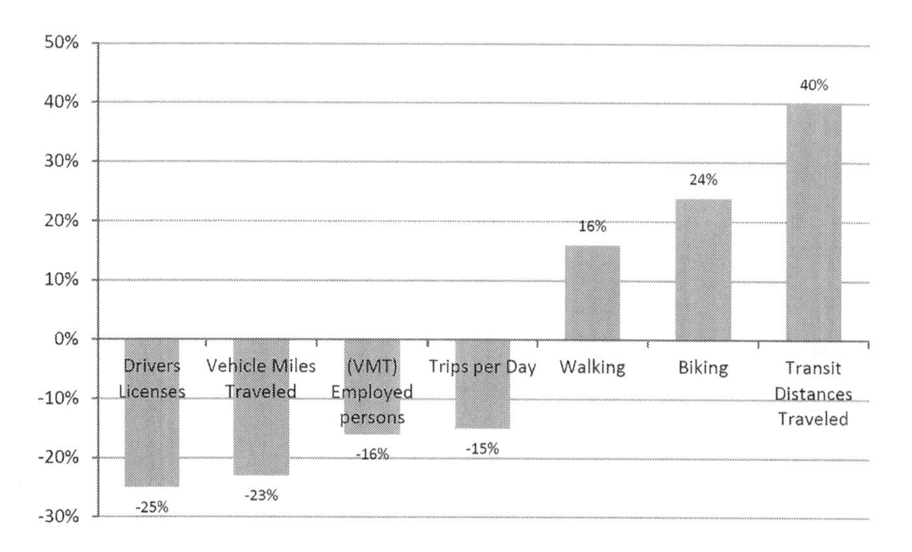

Figure 3.8 Change in Behavior among 16- to 39-Year-Olds, 2001 vs. 2009. (Mid-Region Metropolitan Planning Organization, Albuquerque, NM).

Table 3.8 Transit Mode Shares for Urbanized Areas (from RTPs)

Urbanized Area	*Transit Mode Share Base (%)*	*Transit Mode Share Horizon (%)*	*Transit Mode Share Growth*
Atlanta, GA	1.8	2	11.10%
Austin, TX	2.9	2.7	-6.90%
Baltimore. MD	3.3	3	-9.10%
Boston, MA	6.11	7.21	18.00%
Buffalo, NY	7.00	8	14.30%
Chattanooga, TN	2	3	50.00%
Chicago IL	9	13.5	50.00%
Denver, CO	2.38	2.92	22.70%
Des Moines, IA	7	15	114.30%
Durham, NC	2.8	3.3	17.90%
Flagstaff, AZ	0.32	0.22	-31.30%
Fresno, CA	1.5	1.6	6.70%
Honolulu, HI	5.2	6.1	17.30%
Las Vegas, NV	1.75	1.62	-7.40%
Los Angeles, CA	3	5	66.70%
New York, NY	2.95	3.16	7.10%
Palm Beach, FL	1.6	5	212.50%
Providence, RI	2.8	3.2	14.30%
Provo-Orem, UT	1.4	12	757.10%
Portland, OR	6.9	7.8	13.00%
Raleigh, NC	2.8	3.2	14.30%
Sacramento, CA	2.8	8.2	192.90%
Salt Lake City-Ogden, UT	1.3	1.9	46.20%
San Diego, CA	5.8	11	89.70%
San Francisco-Oakland, CA	16.0	26	62.50%
Seattle, WA	3.1	4.5	45.20%
Tucson, AZ	2.9	2.3	-20.70%
Washington, DC	7	7	0.00%

than is transit use, and sprawl may be depressing walk mode shares in some regions. Another is that unlike transit, expenditures on pedestrian and bicycle facilities are truly modest in most regions.

Comparing Tables 3.8 and 3.9, two facts are worth noting. First, the walk/bike mode share is usually higher than the transit mode share for the same urbanized area. Where it is not, it is likely due to the fact that the mode shares are presented only for work trips, where transit has an edge. Second, transit mode share projections are more often available than walk/

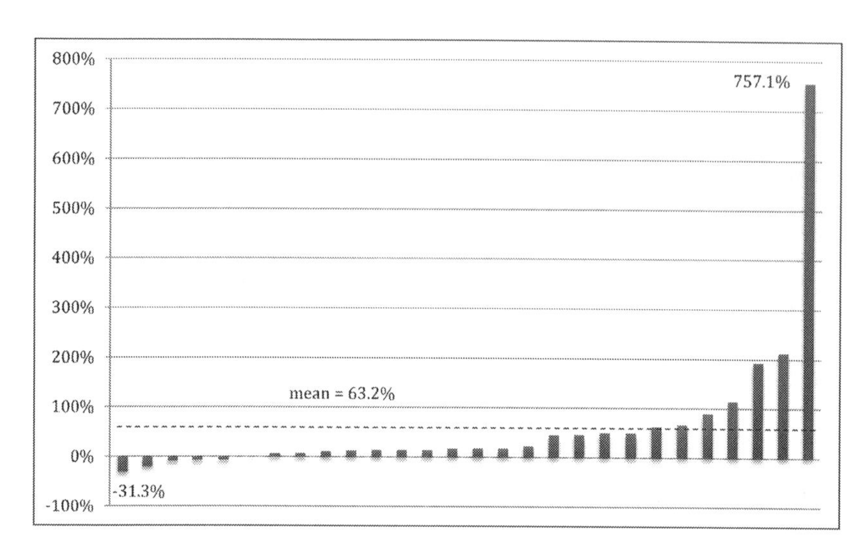

Figure 3.9 Change in Transit Mode Share during Forecast Period.

Table 3.9 Walk/Bike Mode Shares for Urbanized Areas (from RTPs)

Urbanized Area	*Walk/Bike Mode Share Base (%)*	*Walk/Bike Mode Share Horizon (%)*	*Walk/Bike Mode Share Growth*
Austin, TX	5.1	4.9	–3.90%
Boston, MA	14	16	14.30%
Burlington, VT	6.5	8.5	30.80%
Denver, CO	6.99	7.77	11.20%
Des Moines, IA	4	10	150%
Durham, NC	10	10	0.00%
Flagstaff, AZ	10.85	10.4	–4.10%
Fresno, CA	7.3	7.7	5.50%
Honolulu, HI	10.7	11.4	6.50%
Lansing, MI	4.7	6.7	42.60%
Las Vegas, NV	8.24	8.35	1.30%
Palm Beach, FL	2.2	8	263.60%
Portland, OR	4.63	13.9	200.00%
Providence, RI	6.4	6.9	7.80%
Raleigh, NC	10	10	0.00%
Sacramento, CA	3.2	3.8	18.80%
San Diego, CA	2.4	4.8	100.00%
Santa Barbara, CA	4.8	5	4.20%
Seattle, WA	10.2	12.3	20.60%
Tucson, AZ	8.8	7.7	–12.50%
Washington, DC	11	14	27.30%

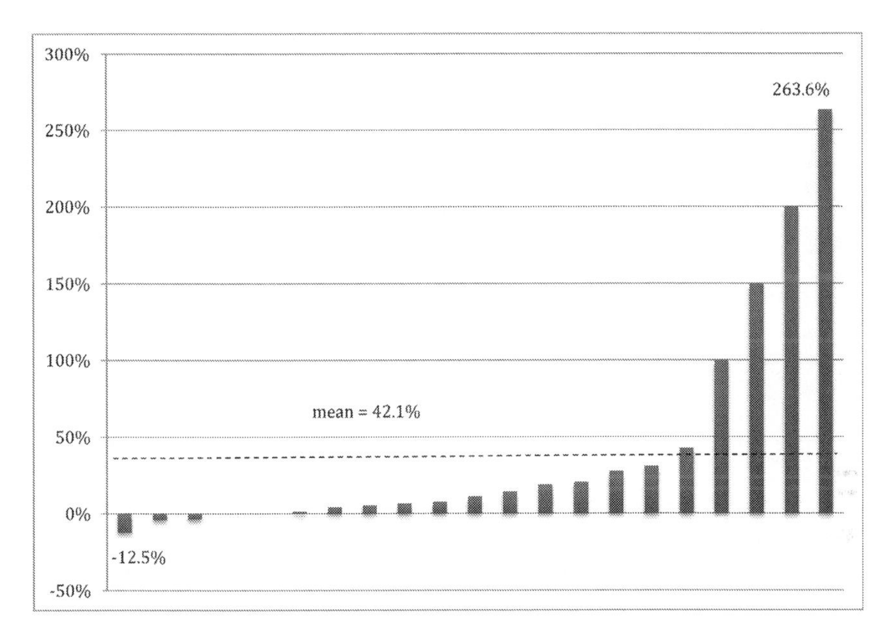

Figure 3.10 Change in Walk/Bike Mode Share during Forecast Period.

bike mode share projections. Transit trips are almost always forecasted in four-step travel demand models. Walk/bike trips often are not. As a mode of travel, walking (and bicycling for that matter) gets less attention than it deserves.

Also worth noting is an implication of these tables. Given the projected shifts to alternative modes, the fact that VMT per capita is growing in most regions must mean that auto trips are getting longer. Auto trips presumably are getting longer because highway capacity improvements and urban sprawl make longer trips possible and necessary (see Chapters 4 and 5).

Expenditures

RTPs plan for millions or even billions of dollars in transportation investments over their planning horizons. Expenditures are sometimes given in today's dollars and other times (per current regulations) given in year-of-expenditure dollars. This accounts for some of the anomalies in the Tables 3.10 through 3.13. Regardless, the amounts invested are always impressive. The smallest expenditure forecast is $236.4 million for the Berkshire County MPO in Massachusetts. The largest is $524,700 million for the Southern California Association of Governments in Los Angeles, CA.

Table 3.10 Total Spending and Highway Spending in Urbanized Areas (from RTPs)

Urbanized Area	*Total Expenditures (in million $)*	*% Freeway Arterial Expansion*	*% Minor Street Expansion*	*% Highway O&M*
Albany, NY	655.39			59.10%
Albuquerque, NM	5,906.46	51.12%		16.71%
Atlanta, GA	79,107	10.98%		27.05%
Austin, TX	28,443	15.18%	17.65%	5.83%
Bakersfield-Kern, CA	11,607.69	56.81%		17.59%
Baltimore, MD	44,907.38	14.94%		53.60%
Berks County, PA	2,154.59	41.68%		
Berkshire County, MA	236.4			
Bismarck, ND	349.42	72.86%		
Boston, MA	69,183			
Bridgeport-Stamford, CT	2,517.63	13.90%		30.60%
Burlington, VT	1,177			64.00%
Charlotte, NC	12,282	33.00%	5.00%	9.00%
Chattanooga, TN	4,928.12	60.20%		
Chicago IL	385,000	33.12%		14.78%
Cincinnati, OH	21,493.13	38.25%		27.53%
Columbus, OH	16,398	19.11%		16.50%
Dallas-Ft. Worth, TX	98,716.5	36.54%	4.45%	10.94%
Dayton, OH	5,654.27	47.76%		
Denver, CO	152,640	6.11%		25.64%
Des Moines, IA	3,282.21			
Durham, NC	7,547	29.40%		13.04%
El Paso, TX	9,352.85			
Flagstaff, AZ	629.52			51.13%
Fresno, CA	4,463.93	39.16%		22.66%
Grand Rapids, MI	6,863.3	44.83%		19.62%
Hartford, CT	2,148.37			40.74%
Honolulu, HI	23,793			54.10%
Houston, TX	57,401.19	50.92%	1.02%	
Kansas City, MO	28,019.1			12.30%
Knoxville, TN		31.46%		61.01%
Lansing, MI	3,917.16	68.15%		8.29%
Las Vegas, NV	9,000	70%		
Little Rock, AR	4,422.9	22.79%		48.57%
Los Angeles, CA	524,700	4.21%		10.81%
Madison, WI	7,653.1	21.13%	42.24%	
McAllen, TX	23,049.88	32.54%		46.06%

Urbanized Area	*Total Expenditures (in million $)*	*% Freeway Arterial Expansion*	*% Minor Street Expansion*	*% Highway O&M*
Miami, FL	35,604	29.05%		4.10%
Milwaukee, WI	19,575	56.15%		9.93%
Minneapolis, MN	83,900	0.83%		12.51%
Newark, NJ	86,861.44	2%		17%
New York, NY	292,333			36.09%
Oklahoma City, OK	7,606.04			86.70%
Philadelphia, PA	52,600	3.04%		7.41%
Portland, ME	506	11.86%	7.91%	9.88%
Portland, OR	22,771.7	56.42%		
Provo-Orem, UT	20,437.6	43.07%	5.69%	11.59%
Raleigh, NC	14,606	39.16%		21.60%
Richmond, VA	7,132.33	20.68%	1.01%	73.90%
Sacramento, CA	49,800	16.76%		36.93%
St. Louis, MO	30,846	12.87%		34.46%
Salt Lake City-Ogden, UT	49,126		21.23%	55.08%
San Antonio, TX	17,624.07	14.17%		2.80%
San Diego, CA	213,826		17.51%	4.60%
San Francisco-Oakland, CA	292,000	5.00%		32.00%
Santa Barbara, CA	7,396.87	31.03%	39.16%	
Seattle, WA	189,300	17.86%		
Spokane, WA	10,863.6			
Stockton, CA	10,950	29.89%		35.39%
Tampa-St. Petersburg, FL	17,058	7.60%		26.9%
Tucson, AZ	18,312.16	66.94%		
Tulsa, OK	4,138.5	41.53%		23.21%
Virginia Beach, VA	26,900	89.22%		
Washington, DC	244,000	11.07%		29.51%
Worcester, MA	4,394.32	56.56%		
Youngstown, OH	1,101.8	85.02%		

Some plans provide only project lists with estimated costs. Most offer some sort of summary by category of expenditure. Line items differ from plan to plan, which accounts for the different samples in these tables.

What is apparent is the following:

MPOs continue to devote significant revenues to highway (freeway and principal arterial) expansion, despite the induced travel and induced development that follow from these investments. However, there is tremendous variation across RTPs, from a low of 0.83 percent of total expenditures in Minneapolis-St. Paul, MN to a high of 89.2 percent in Virginia Beach, VA. For RTPs that break out freeway/arterial capital expenditures as line items, the mean value is 33.6 percent, whereas the median value is 31.5 percent.

Most MPOs do not earmark expenditures for minor roads (minor arterials, collectors, and locals). For RTPs that break out minor street expenditures as a line item, the mean value is 14.8 percent, and the median is 7.9 percent.

MPOs are uneven in their funding of highway preservation, maintenance, and operations. The lowest share of total funds is 2.8 percent in San Antonio, and the highest is 86.7 percent in Oklahoma City, OK. For those that break out expenditures, the mean value is 28.5 percent, and the median value is 23.2 percent. It is not clear whether more is better, as it may simply reflect lack of preventative maintenance in the past, which has finally caught up with the region.

Most MPOs devote substantial resources to transit (see Table 3.11). Transit receives about a third of total funds in many large regions, and more than half in a few. The highest share of funds devoted to transit is 69.3 percent in Philadelphia. For RTPs that earmark funds, the mean value is 32.2 percent, the median value is 32.0 percent. In contrast, most MPOs spend relatively little on pedestrian and bicycle facilities, considerably less than the share of trips by these modes. For RTPs that earmark funds, the mean expenditure on pedestrian/bicycle facilities is 2.6 percent of total funds, whereas the median is 2.3 percent. Many MPOs treat such facilities as local rather than regional responsibilities and do not earmark. Also, they do not break out sidewalk and bike lane costs that are part of larger highway projects.

Likewise, MPOs spend only small shares of total funds on transportation system management (TSM), intelligence transportation systems (ITS), and travel demand management (TDM), even though these are declared to be cost-effective and priorities in many RTPs (see Chapter 6). The highest percentage of total funds spent on TSM is 35 percent in Columbus, OH. It is an outlier. For RTPs with line item detail, the mean spending on TSM is 5.5 percent of total funds, and the median is 2.3 percent (see Table 3.12). The highest percentage of total funds spent on TDM is 3.4 percent in Dallas-Ft. Worth, TX; for RTPs with line item detail, the mean is 0.9 percent and the median is 0.6 percent.

One final category of spending is noteworthy. Recently, some MPOs have begun to earmark a small percentage of total funds to land planning/smart growth, making grants to local governments for these purposes. In 1998, San Francisco began its Transportation for Livable Communities Program and has since awarded over $200 million to support various projects linking land use and transportation. Atlanta followed suit in 1999 with the Livable Centers Initiative, and the practice has since spread to quite a few other regions of the United States. The amounts are still miniscule but the trend is promising.

Table 3.11 Total Spending and Spending on Alternative Modes in Urbanized Areas (from RTPs)

Urbanized Areas	*Total Expenditures (in million $)*	*% Transit Capital*	*% Transit O&M*	*% Walk/ Bike Facilities*
Albany, NY	655.39		12.20%	0.56%
Albuquerque, NM	5,906.46	18.24%		4.35%
Atlanta, GA	79,107	8.04%	39.22%	2.34%
Austin, TX	28,443	10.28%	36.09%	1.56%
Bakersfield-Kern, CA	11,607.69	20.77%	15.92%	6.50%
Baltimore, MD	44,907.38	9.67%		0.21%
Berks County, PA	2,154.59	18.21%		
Berkshire County, MA	236.4			
Bismarck, ND	349.42	19.62%		7.52%
Boston, MA	69183			0.23%
Bridgeport-Stamford, CT	2,517.63	15.10%	25.80%	2.80%
Burlington, VT	1,177	8.00%		3.30%
Charlotte, NC	12,282	25.00%	27.00%	
Chattanooga, TN	4,928.12	14.56%		0.50%
Chicago IL	385,000	8.21%	41.17%	
Cincinnati, OH	21,493.13	5.09%	28.75%	0.01%
Columbus, OH	16,398	26.67%		1.37%
Dallas-Ft. Worth, TX	98,716.5	14.37%	18.36%	1.52%
Dayton, OH	5,654.27	51.33%		0.33%
Denver, CO	152,640	7.11%	15.44	0.65%
Des Moines, IA	3,282.21	17.90%		
Durham, NC	7,547	35.34%	18.21%	2.39%
El Paso, TX	9,352.85			
Flagstaff, AZ	629.52	22.07%		5.99%
Fresno, CA	4,463.93	35.66%		2.52%
Grand Rapids, MI	6,863.3	5.68%	17.87%	1.17%
Hartford, CT	2,148.37	55.07%		2.32%
Honolulu, HI	23,793	28.00%		
Houston, TX	57,401.19	3.77%	43.28%	0.51%

(*Continued*)

Table 3.11 (Continued)

Urbanized Areas	*Total Expenditures (in million $)*	*% Transit Capital*	*% Transit O&M*	*% Walk/ Bike Facilities*
Kansas City, MO	28,019.1		14.10%	
Knoxville, TN	2,153.53	4.57%		1.26%
Lansing, MI	3,917.16	5.54%	8.83%	1.78%
Las Vegas, NV	9,000	19%		3%
Little Rock, AR	4,422.9	3.34%	24.74%	0.57%
Los Angeles, CA	524,700	19.29%	26.55%	1.14%
Madison, WI	7,653.1	6.93%	25.59%	1.18%
McAllen, TX	23,049.88	9.14%		2.93%
Miami, FL	35,604	14.47%	51.54%	0.28%
Milwaukee, WI	675	4.74%	29.19%	
Minneapolis, MN	83,900	9.15%	25.19%	
Newark, NJ	86,861.44	11%	39%	
New York, NY	292,333		54.99%	
Oklahoma City, OK	7,606.04		10.00%	3.30%
Philadelphia, PA	52,600	32.39%	36.90%	1.00%
Portland, ME	506	3.95%		2.96%
Portland, OR	22,771.7	28.18%		10.60%
Provo-Orem, UT	20,437.6	21.45%	10.41%	
Raleigh, NC	14,606	22.29%	9.18%	2.20%
Richmond, VA	7,132.33	4.41%		
Sacramento, CA	49,800	9.58%	22.35%	8.03%
Salt Lake City-Ogden, UT	49,126	19.28%	4.40%	
San Antonio, TX	17,199.07	44.09%	7.91%	0.38%
San Diego, CA	213,826	15.44%	24.65%	1.24%
San Francisco-Oakland, CA	292,000		55%	
Santa Barbara, CA	7,396.87	26.37%		2.72%
Seattle, WA	189,300	51.35%		0.16%
Spokane, WA	10,863.6			
St. Louis, MO	30,846		52.68%	
Stockton, CA	10,950	32.15%		2.58%
Tampa-St. Petersburg, FL	17,058		39.40%	2.30%
Tucson, AZ	18,312.16	27.85%		2.93%
Tulsa, OK	4,138.5	13.41%	19.23%	2.61%
Virginia Beach, VA	26,900	8.18%		2.60%
Washington, DC	244,000	6.15%	53.28%	
Worcester, MA	4,394.32	29.86%	10.62%	11.96%
Youngstown, OH	1,101.8	12.35%		2.06%

Table 3.12 Total Spending and Spending on Transportation Management in Urbanized Areas (from RTPs)

Urbanized Area	*Total Expenditures (in million $)*	*% TSM/ITS*	*% TDM*
Albany, NY	655.39	2.33%	
Albuquerque, NM	5,906.46	3.29%	0.60%
Atlanta, GA	79,107		0.62%
Austin, TX	28,443	3.83%	
Baltimore, MD	44,907.38		0.63%
Bridgeport-Stamford, CT	2,517.63	8.40%	
Burlington, VT	1,177	12.50%	
Charlotte, NC	12,282		1.00%
Cincinnati, OH	21,493.13	0.30%	
Columbus, OH	16,398	35.03%	
Dallas-Ft. Worth, TX	98,716.5		3.38%
Denver, CO	152,640	1.62%	0.07%
Durham, NC	7,547	1.23%	0.40%
El Paso, TX	9,352.85		
Las Vegas, NV	9,000	8%	
Los Angeles, CA	524,700	1.30%	0.76%
Newark, NJ	86,861.44	8.44%	
Portland, OR	22,771.7		1.17%
Raleigh, NC	14,606	2.15%	
San Diego, CA	213,826		0.53%
Santa Barbara, CA	7,396.87	0.06%	0.65%
Seattle, WA	189,300	0.74%	0.79%
Spokane, WA	10,863.6		
Tampa-St. Petersburg, FL	17,058	3.70%	
Worcester, MA	4,394.32	0.15%	

Conclusion

Regional transportation plans are fundamental documents for metropolitan planning organizations as they show the transportation vision for the region and outline transportation projects that are vital to growth and economic development. Although there is significant variation, trends in population growth, employment, mode share, and expenditures are enlightening. Despite a loss of population in three MPOs and a loss of employment in one MPO, we still see an increase in VMT across the board. In many cases, the growth in VMT is faster than the population growth for the urban area.

As noted in Chapter 2, regional transportation planning historically has focused on mobility. However, nearly all of the plans we reviewed reflected new, nonmobility-related priorities, such as coping with climate change in the future, environmental justice, and sustainable growth. Projected increases in walk/bike and transit mode shares also show changing priorities of communities. Although the amount of funding dedicated to walking and biking infrastructure is no more than a small percentage of the overall budget for all MPOs that reported these data, walk and bike mode shares are expected to increase, in some cases significantly, for 16 of the 21 MPOs. Transit operations and maintenance, meanwhile, represent a significant share of the budget in many plans, and we see an increase in projected transit mode share for all but six of 28 MPOs.

This quantitative analysis shows where regions are projected to go in terms of population, employment, and VMT growth and where they will spend their money, but the process by which they plan and grow is also critical and is the subject of later chapters.

4 Quantitative analysis of regional scenario plans

Introduction

One of the leading trends in metropolitan transportation planning processes that has emerged since the early 1990s is a style of regional planning that seeks to integrate land use and transportation functions and outcomes. The motivation behind this integration was a desire by planners to create a process more accurately reflecting the reality of people's daily lives. Because most people when they drive, or bus, or bike around town are trying to go someplace to get something done, planners opined, shouldn't planning processes take into account not only the transportation links between here and there, but also the locations and functions of both here and there? To facilitate the creation of such a planning process, planners borrowed from methods developed by business and military strategic planners collectively known as scenario analysis.

Scenario analysis: asking "what if"?

Scenario analysis begins with the recognition that the future is fundamentally unknowable and unpredictable. We may have some guesses about how the future might unfold in some general way – based on logical interpretations of past chains of events – but we can never be certain. Making a single prediction about how things might turn out seems as fraught with uncertainty as the seven-day weather forecast on the evening news.

Still, as with the weather forecast, we can look back and see how a set of conditions in the past brought about particular results; we can see how certain factors, operating in certain ways, might lead to specific outcomes. If, say, under XYZ conditions X goes a particular way, then we might expect A to occur, or if Y goes a particular way we might expect B to be the outcome, and so forth. Our problem (or, rather, one of our problems) is that we are not certain which direction X, Y, or any other variable might go and we are not sure whether we have correctly identified all of the important variables. What if some completely unforeseeable event intercedes (in SimCity talk, what if Godzilla attacks)?

Scenario analysis allows us to reduce uncertainty about the future – to frame, or bound, that uncertainty – by broadening our thinking beyond a single predicted outcome. Logically, if instead of targeting one (necessarily uncertain) projection as the basis for planning we use a series of possible outcomes, we increase the odds of having plans that are responsive to what actually occurs.

The bounding of uncertainty does not eliminate uncertainty, however, and scenarios are not and should not be considered predictive. In the words of Michael Porter, one of the leading thinkers about business-style scenario planning, a scenario is not a forecast, but one possible future outcome – "an internally consistent view of what the future might turn out to be" (Porter, 1985, p. 446). To reinforce this characteristic of scenarios, many planners refer to them as stories – stories about the future. Labeling something a story signals, at least in English vernacular, that the information is to be understood as illustrating some underlying truth about a situation, without necessarily being 100 percent accurate.

The story-based definition of *scenario* is reflected in the word's etymology. Derived from Italian, the word originally referred to the scenery or set used in theatrical productions. The theatrical associations continued in English with the first widespread use of the term coming from Hollywood, where screenwriters used (and still use) scenario to refer to the verbal or graphic sketches that frame the plot, characters, situations, camera directions, and so on for individual movie scenes. Given these connections between scenarios and stories, an analysis using multiple scenarios – a scenario planning process – can be properly understood as one that provides a series of stories about the future that collectively define a range of possible outcomes.

The roots of scenario analysis lay within the broader topic of adaptive response technique, the military applications of which can be traced back centuries, at least as far as Sun Tzu's famous 6th century BCE treatise, the Art of War (Giles, 1910). More modern applications include those done by the RAND Corporation, where during the 1950s scenario analysis was used to anticipate, and prepare for, possible Soviet nuclear attack strategies (Kahn, 1962). The apocryphal business application was Royal Dutch/Shell's use of scenario analysis to effectively anticipate the OPEC Oil Embargo of 1973 (Schwartz, 1991). Since then, scenario analysis has become fairly common in business circles and the business-based literature is well developed.

Scenario analysis came to transportation planning comparatively late. It was preceded by a similar, although much more limited, technique called alternatives analysis, which arose in the early 1960s in response to the Federal Aid Highway Act's mandate that metropolitan regions conduct "3C" (continuing, cooperative, comprehensive) planning. By the 1970s, alternatives analysis had become a standard component of most project-level environmental studies, thanks largely to the passage of the National Environmental Policy Act in 1969 (Weiner, 1999). Alternatives analysis bears some obvious similarities to scenario planning, chiefly in the consideration of alternative responses to an identified problem. Its practice, however, has in most cases been highly constrained, allowing consideration of alternate transportation improvements, but excluding other possible variables, such as land use patterns and economic conditions and policies.

Frustration with the narrowness – and artificiality – of this approach led to the development of two projects in the late 1980s/early 1990s that used land use patterns as a variable between alternatives, in addition to possible transportation improvements. The two projects – Montgomery County's Comprehensive Growth Policy Study (1989) and 1000 Friend of Oregon's LUTRAQ project (Bartholomew, 1995) – were not the first explorations into using land use as a variable in a transportation analysis. Most of the earlier examples, however, were academic in origin (e.g., Edwards & Schofer, 1976; Mazziotti et al., 1977; Peskin & Schofer, 1977). What set the Montgomery County and 1000 Friends studies apart and arguably made them vanguard projects is that they were done in decision-making contexts that resulted in the adoption and implementation of land use planning and infrastructure investment policies.

Land use-transportation scenario planning in America

The two projects helped foster a number of similar studies across the United States. An analysis from 2005 shows there were 80 such projects completed between 1989 and 2003 (Bartholomew, 2007a), and a follow-up study in 2010 revealed an additional 28 (Figure 4.1).

Though these studies reveal a great deal of variation in planning practices (which can be observed in two annotated bibliographies of the projects; see www.fhwa.dot.gov/planning/scenplan/practices.htm), they all tend to follow a fairly consistent work program. The modal project begins with the assessment of a trend scenario where urban development and transportation investment patterns of the recent past are assumed to continue to the planning horizon (typically 20 to 50 years in the future). Analysis of this scenario's impacts on land consumption, air quality, and transportation systems almost always indicates results that are dismal, often shocking. The trend scenario, thus, not only creates a baseline for comparing with other, alternative scenarios, but frequently also provides political motivation for crafting alternative scenarios. The alternative scenarios are then assessed for their impacts on the same measures, and the results are compared to help inform the development of new policies on transportation, land use, and related topics.

This storyline falls neatly into the structure of rational theory (sometimes called rational-comprehensive theory), a familiar construct in planning circles. Its step-wise, seemingly objective form speaks intuitively to human problem-solving processes. One begins by articulating a problem, then proceeds to create, test, and compare a series of alternatives for solving that problem, selecting the option that performs best (or most optimally)

Figure 4.1 U.S. Land Use-Transportation Scenario Planning Projects, 1989–2010.

(Anderson, 2003). The critics of rational-comprehensive theory are legion, and its application to understanding scenario planning practice has many weak points (see Bartholomew, 2005, 2007a). It is, nevertheless, useful for seeing the general flow of scenario planning processes.

The Region 2040 project sponsored by Metro in Portland, Oregon, provides a handy illustration. Metro, the Portland-area regional government, was tasked by the state legislature in the 1970s to establish and administer the region's urban growth boundary. In the late 1980s/early 1990s Metro was facing pressures to, on the one hand, expand the boundary for economic development purposes and, on the other, maintain the boundary to rein in sprawl, protect open spaces, and support transit use. Metro initiated the Region 2040 project to address the seeming conflict in mandates and began by drafting a general vision for the region's future and updating a series of urban growth goals and objectives. These documents then drove the creation of four scenarios – a trend scenario plus three alternatives – each of which accommodated expected future growth through 2040. The scenarios were then analyzed quantitatively and qualitatively, and the results were arrayed to facilitate comparisons across scenarios. After extensive public outreach and feedback, Metro adopted a Region 2040 Growth Concept scenario that blended features of the three alternative scenarios and developed a series of implementing policies.

Though the Region 2040 project dates from the mid-1990s, most of the projects completed since then have followed a similar arc and produced results that are similar in direction if not degree.

Scenario inputs

If scenario planning facilitates telling stories about the future, what kinds of stories do land use-transportation scenario projects tell? As with Region 2040, most land use-transportation scenario projects are inspired by concerns about sprawl, open space, and transportation (Bartholomew, 2007a). Hence, the stories they tell tend to reflect futures where development patterns are more compact (i.e., have a smaller footprint) than their respective trend counterparts. The 23 scenario projects depicted in Figure 4.1 that were completed between 2004 and 2010 contain 127 alternative scenarios, 115 with data about land consumption. Of those, only 3 scenarios are more consumptive and 17 as consumptive as their respective trend scenarios. The remaining 107 scenarios consume from 1 to 65 percent less land than the trend (Figure 4.2).

Accommodating the same population with less land means, of course, that development densities would need to be higher. This is illustrated in Figure 4.3, where the densities of alternative scenarios increase by as much as 2.8 times that of the trend scenarios.

Many scenarios include other, related land use features in addition to compactness/density. One common attribute is the so-called mixing of land

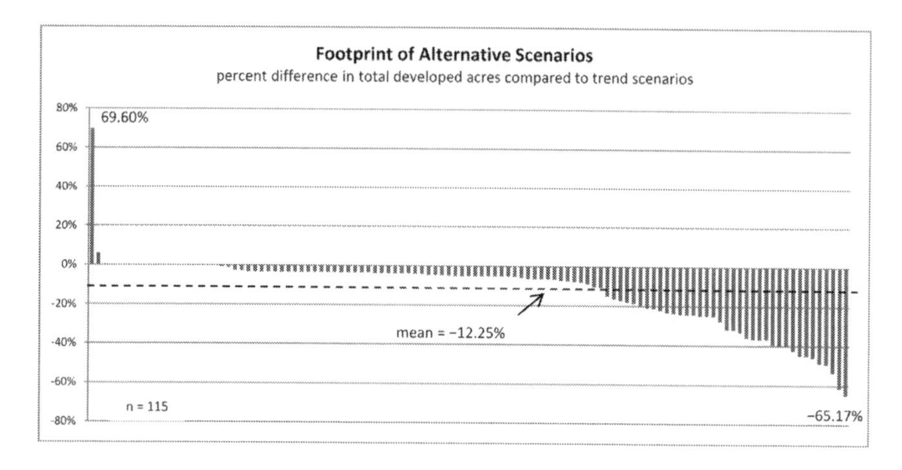

Figure 4.2 Percentage Difference in Developed Acres: Alternative vs. Trend Scenarios.

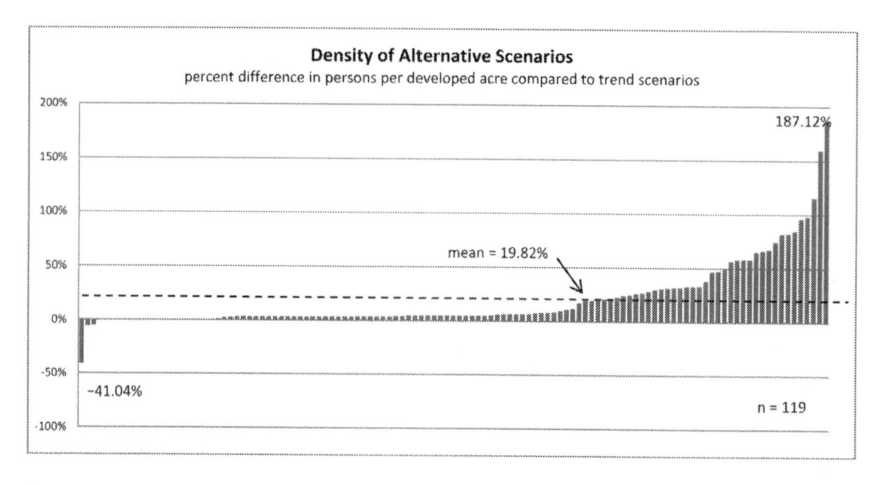

Figure 4.3 Percentage Difference in Persons per Developed Acre: Alternative vs. Trend Scenarios.

uses – integrating shops, offices, houses, and apartments into the same neighborhood, for example. This feature stands in contrast to the conventional suburban neighborhood where uses are commonly separated into their own enclaves. Another common scenario feature is increasing easy access to transit stations or stops (Figure 4.4). All of these land use factors imply an additional scenario trait that is not frequently mentioned – destination accessibility – which refers to the degree to which access to the daily needs of life can be reached with relative ease and convenience. The inclusion of many of these land use elements is driven, in part, by a desire to reduce

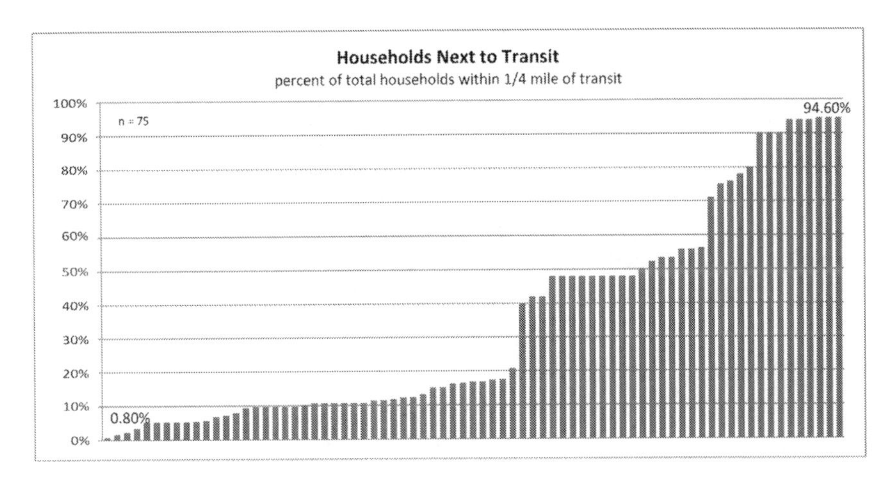

Figure 4.4 Percent of Total Households within ¼ Mile of Transit.

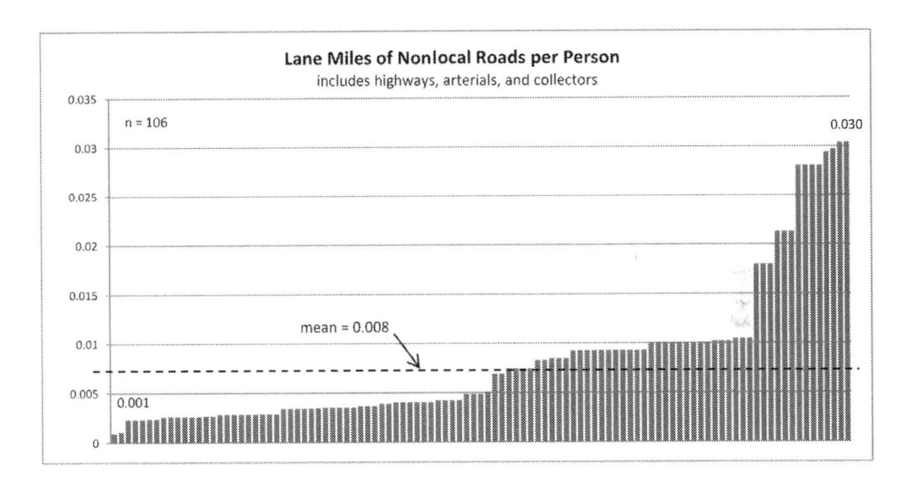

Figure 4.5 Lane Miles of Highways, Arterials, and Collectors per Person.

automobile use and an understanding that dense, mixed-use development patterns with good assess to transit and destinations are less auto-reliant (Ewing & Cervero, 2010).

Given that the sponsors of scenario planning projects are frequently regional transportation agencies (known as metropolitan planning organizations or MPOs), it is not surprising that, in addition to land use features, most land use-transportation scenario projects vary the amount and type of transportation improvements between scenarios. Some scenarios may favor expansion of regional arterials and highways (Figure 4.5), whereas others may focus on additions to local transit systems (Figure 4.6). Only a handful incorporate improvements to pedestrian and bicycle networks or policies to increase the cost of driving, for example, by charging higher parking fees or roadway tolls.

An important factor underlying all scenarios is the amount of growth they are intended to accommodate. The land use-transportation scenario projects completed between 2004 and 2010 show variation not only in the length of time horizons studied (from 20 years in Washington, DC, to 48 years in Missoula, MT), but also in the annual rates of growth assumed (from 0.24 percent in Lansing, MI to 3.77 percent in St. George, UT). A composite measure incorporating both the time horizon length and the annual growth rate (Figure 4.7) shows a broad range in the percentage of growth being accommodated by the scenarios, from just under 7 percent to more than 180 percent.

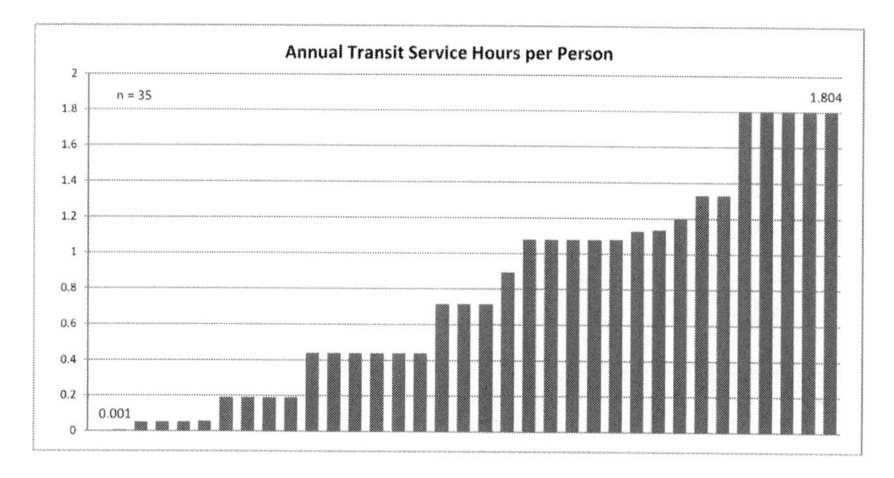

Figure 4.6 Annual Transit Revenue Hours of Service per Person.

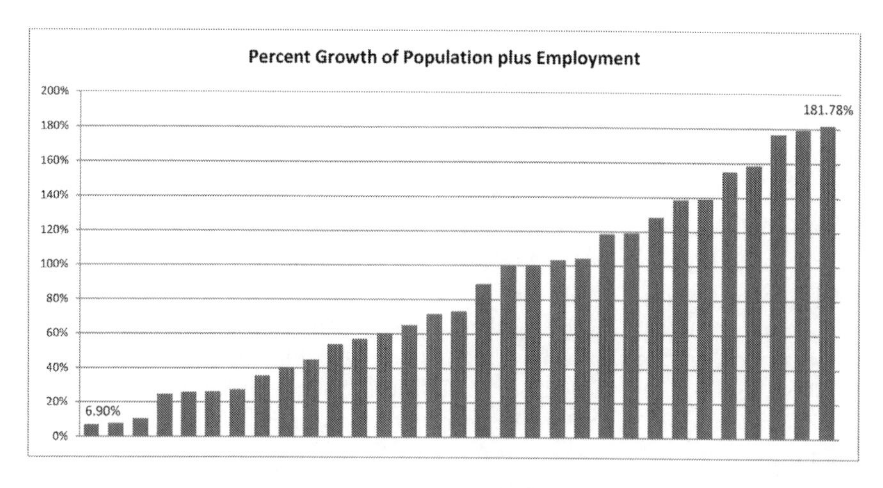

Figure 4.7 Percent Change in Population plus Employment for All Scenario Projects.

Analytical tools

As with stories, the meaning of scenarios is revealed in their outcomes, and scenario outcomes are generally determined using analytical tools to assess scenario impacts on important outcomes. Because the sponsors of land use-transportation scenario projects are usually MPOs, the most frequently used assessment tools for this type of scenario analysis are computer-based models that estimate transportation demand and travel patterns.

With one exception, all of the models used in the scenario projects completed between 2004 and 2010 were some version of the standard Urban Transportation Modeling System (UTMS). UTMS models, initially created to estimate demand for large-scale highway facilities (Pas, 1995), are insensitive to the impacts of land use variables (except, to a degree, destination accessibility) on travel mode and trip length. A recent survey of local and regional governments in California conducted by DKS Associates and University of California (2007) identified a number of features in current UTMS-based modeling structures and practices that limit their ability to evaluate the land use impacts of scenarios. These features include the inability to model trip chaining behavior; the neglect of walk and bike trips; the use of fixed vehicle trip rates by land use type; the failure to consider the effect of building, street, and sidewalk layout; the use of large travel analysis zones that blur land use patterns; and the failure of transportation system performance to feedback to land use allocation decisions. These shortcomings are echoed in other recent critiques of modeling systems and practices (Beimborn, Kennedy, & Schaefer, n.d.; Cervero, 2006; Committee for Determination of the State of the Practice in Metropolitan Area Travel Forecasting, Transportation Research Board [TRB], 2007; Johnston, 2004; Walters, Ewing, & Schroeer, 2000).

Modeling capacity varies widely between metropolitan regions, however, and a survey of the agencies responsible for the 2004–2010 land use-transportation scenario projects demonstrates this. The survey, based on the DKS study, ordinally ranked modeling functions according to their relative importance in measuring the impacts of land use variables (Figure 4.8). Planners at the sponsoring agencies – the survey respondents – were directed to indicate which of those components was part of the modeling system used for their scenario project.

The results of the survey, displayed in Table 4.1, range from the Cheyenne, Wyoming, model, which contains only the first two elements, to the Puget Sound Regional Council model, which contains all of the elements listed (except post-processing, which is probably unnecessary given the other components of the agency's modeling system). The progression from the most basic elements, on the left of the table, to the more advanced components, on the right, does not rigorously follow the progression depicted in Figure 4.8 – there are frequent gaps in the middle of many of the responses. Nevertheless, the table does give an overall sense of evolution, from left to right, which is at the root of the DKS study.

Table 4.1 Results of Scenario Planning Model Survey

	Daily Vehicle Trip Model	*Peak & Daily Travel*	*Transit/ Highway Mode Choice*	*Transit Network & Daily Assignment*	*Supply & Demand Equilibration*	*Income Strat. in Dist. & Mode Choice*	*Auto Ownership Sensitive to Land Use*	*Travel Time Feedback*	*Ped/Bike in Mode Choice*	*Multiple Modes of Access to Transit*	*Distrib. Sensitive to Multimodes*	*Disaggregate Simulation of HHs*	*Ped/Bike Networks*	*Activity- and Tour-Based Modeling*	*Post-Processing of D Variables*	*Integrated Land Use-Transportation Model*
San Luis Obispo	X	X													X	
San Francisco Bay Area	X	X	X	X	X	X	X		X	X			X			
SJVB: San Joaquin[1]	X	X	X					X								
SJVB: Stanislaus	X	X	X			X	X	X								
SJVB: Merced	X				X											
SJVB: Madera	X				X			X				X				
SJVB: Fresno	X	X	X	X		X		X	X	X					X	
SJVB: Tulare	X	X			X			X								
SJVB: Kings	X	X			X			X								
SJVB: Kern	X	X	X	X		X		X	X	X	X				X	X
Shasta	X	X	X	X					X		X					
Wilmington, DE	X	X	X	X				X		X		X				
Washington, DC	X	X	X		X		X	X		X	X					
Central Florida	X		X	X	X					X		X				
Atlanta	X	X	X	X	X	X	X	X		X	X	X		X	X	X
Boise, ID	X	X		X					X	X					X	
Chicago	X	X	X	X	X		X	X		X	X	X			X	

Champaign, IL	X		X		X			X							
Lansing, MI	X	X	X	X			X	X		X					
Missoula, MT	X	X	X			X	X								
Albany, NY	X	X						X							
Binghamton, NY		X			X		X	X							
Durham, NC	X	X	X	X	X	X		X		X	X	X			
Eugene, OR	X				X			X				X			X
Brownsville, TX	X	X				X		X				X			
Waco, TX	X				X							X			
Logan, UT	X	X	X	X	X	X	X	X	X	X		X			
St. George, UT	X		X							X					
Salt Lake City, UT	X	X	X	X	X	X	X	X	X	X					
Puget Sound, WA	X	X	X	X	X	X	X	X	X	X	X	X	X	X	X
Cheyenne, WY	X	X													

[1] SLVB refers to the San Joaquin Valley Blueprint project, which utilized the individual travel demand models of the eight MPOs participating in the project.

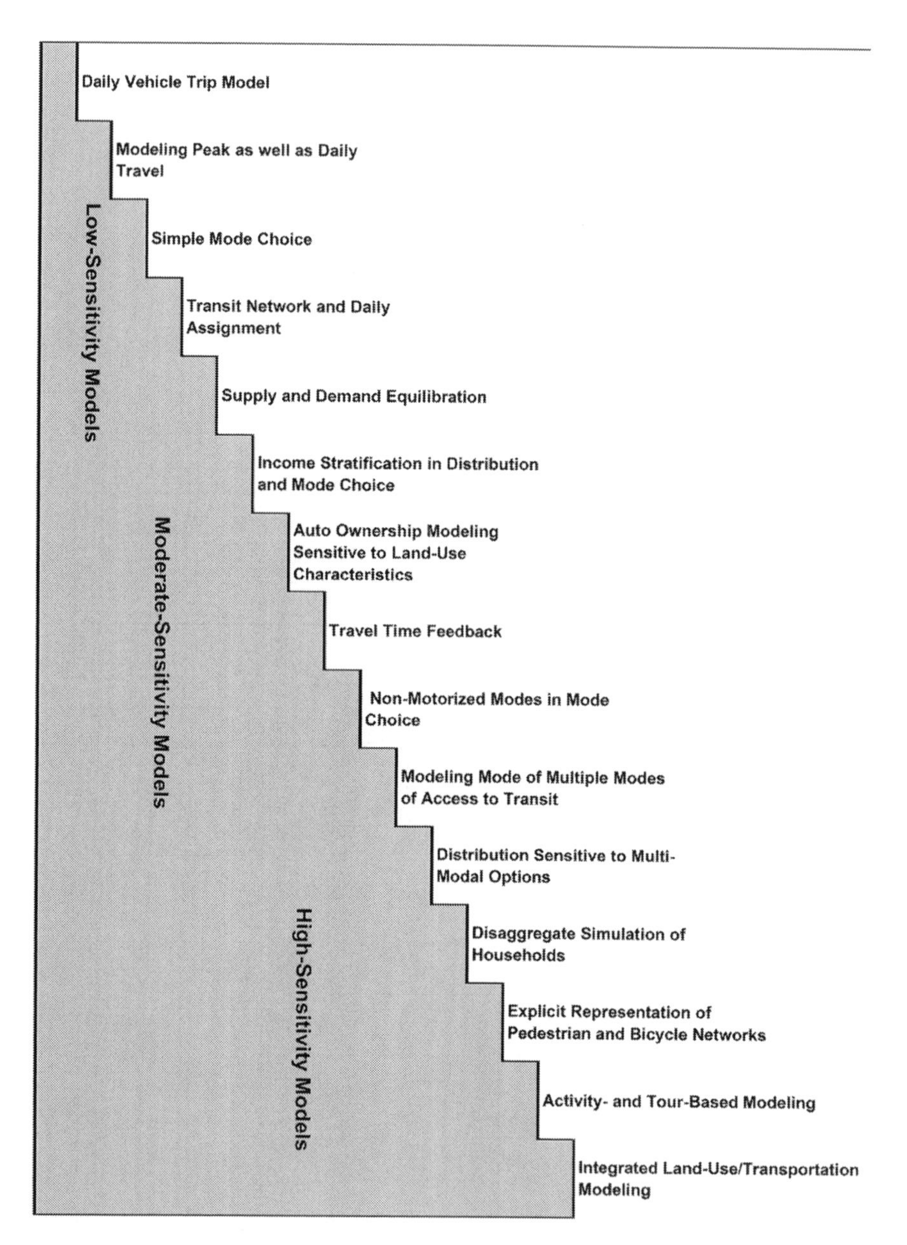

Figure 4.8 Model Functions Ranked in Order of Importance to Assessing Land Use Scenarios.

(DKS, 2007, p. E-5).

Scenario outputs

Naturally, the dominance of transportation-based modeling systems in scenario assessment means that the primary output data from land use-transportation scenario projects is transportation related. The most common transportation measure, and the one that probably best captures overall travel demand, is vehicle miles traveled (VMT), which is the sum of all the miles traveled by all of the vehicles within a specified area during a specified time. The range in daily VMT per person among the 150 scenarios – both alternative and trend – contained in the 2004–2010 scenario projects is quite broad, running from 12 to 63 vehicle miles (Figure 4.9). The percentage difference in VMT between alternative scenarios and their respective trend scenarios is also broad, running from +24 percent to –30 percent (Figure 4.10).

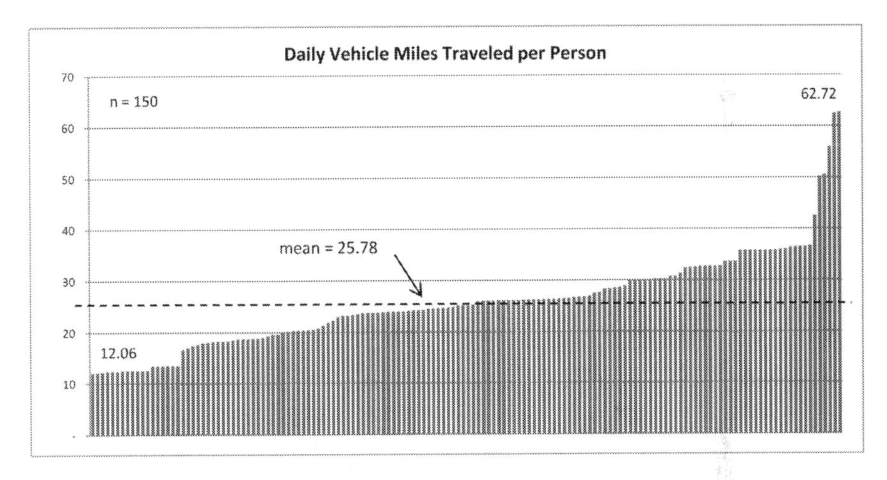

Figure 4.9 Daily VMT per Person.

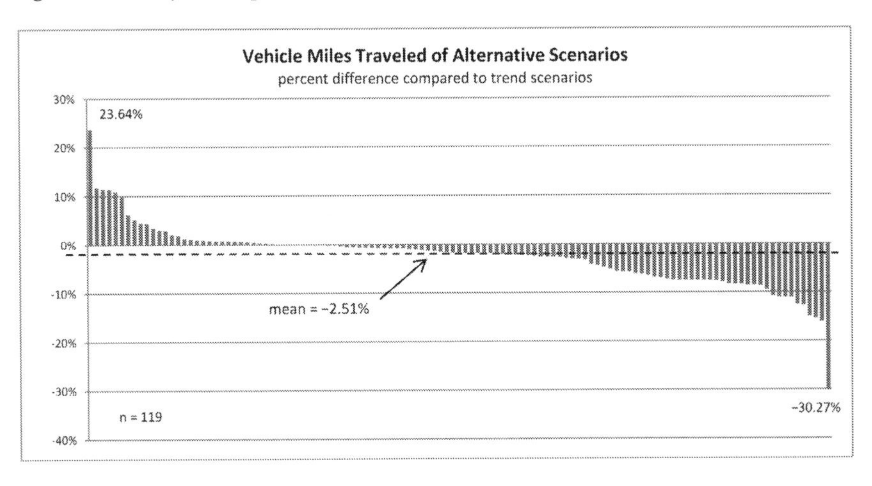

Figure 4.10 Percent Difference in VMT: Alternative vs. Trend Scenarios.

Bivariate analysis

Analysis of what scenario elements are associated with what levels of VMT begins with so-called bivariate analysis – comparing one variable against another. Comparison of density (as measured by persons per acre) and VMT per person (Figure 4.11) shows the data points moving in a consistent direction – as density increases, VMT decreases. The scatter of points off the trend line, however, indicates that other factors in addition to density are influencing VMT levels. In other words, whereas density is an important influence on travel demand, its effect is only partial, a finding consistent with much of the land use-transportation literature (e.g., Ewing & Cervero, 2001).

Bivariate analysis of the amount of transportation improvements included in scenarios also provides only a partial explanation for the variation in VMT. Scatterplots of road lane miles versus VMT (Figure 4.12) and transit

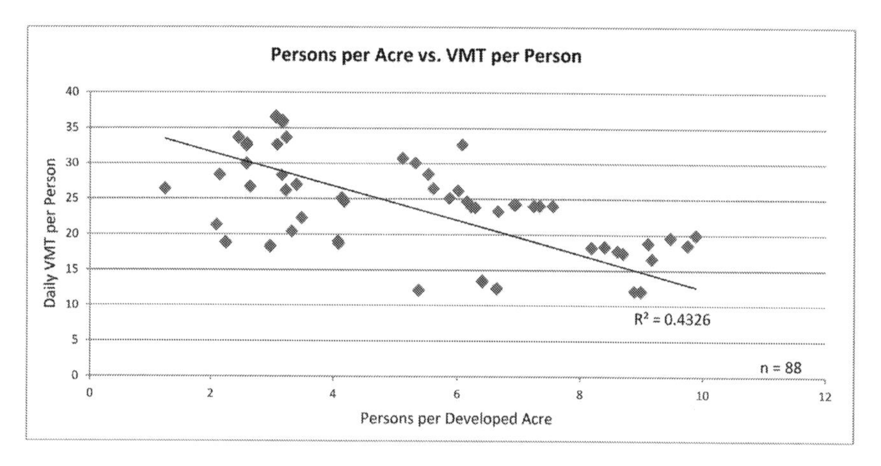

Figure 4.11 Scatterplot of Persons per Acre and Vehicle Miles Traveled per Person.

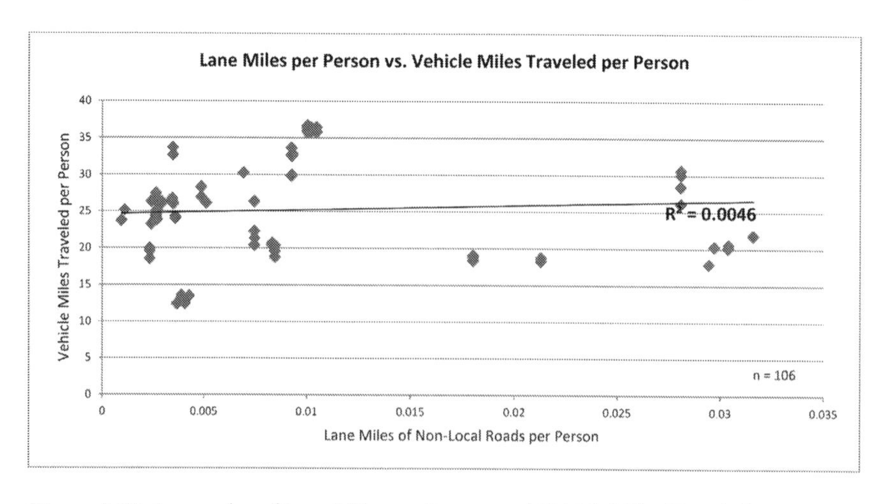

Figure 4.12 Scatterplot of Lane Miles per Person and Vehicle Miles Traveled per Person.

service hours versus VMT (Figure 4.13) show that VMT increases with additional road capacity and decreases with additional transit service, findings that are consistent with research on the "induced demand" effects between transportation supply and travel consumption (e.g., Cervero, 2002). The scatter around these trend lines, however, is even greater than in the density scatterplot, suggesting that the relationships are not as strong.

The percentage of housing within walking distance of transit, which reflects both on land use patterns and transit supply, shows a stronger relationship (Figure 4.14). As transit availability increases, this graph suggests, the amount of driving by residents declines.

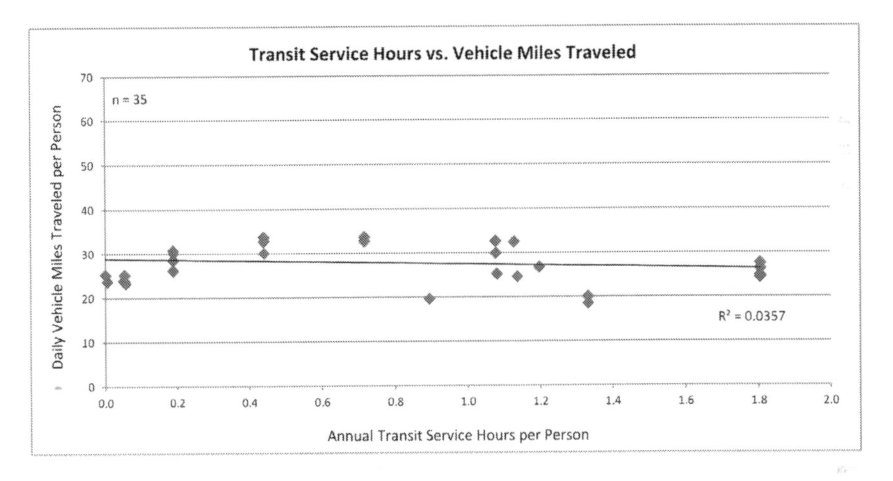

Figure 4.13 Scatterplot of Annual Transit Revenue Hours per Person and Vehicle Miles Traveled per Capita.

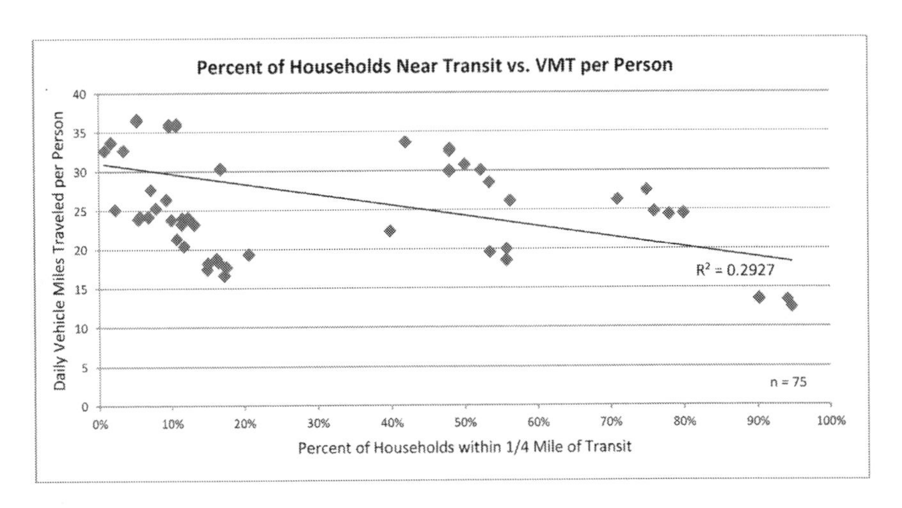

Figure 4.14 Scatterplot of Percent of Households Near Transit and Vehicle Miles Traveled per Person.

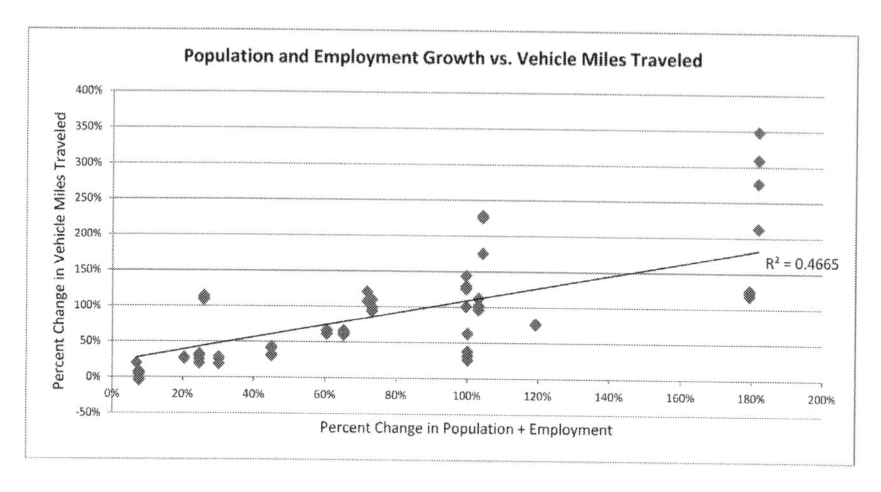

Figure 4.15 Scatterplot of Percent Change in Population + Employment and Percent Change in VMT.

Not surprisingly, the greater the growth in population and employment, the greater the growth in VMT (Figure 4.15). As with the other scatterplots, this graph indicates that many factors other than planning horizons and growth rates are also in play. What is unusual about this graph are the clusters of data points with the same growth rate but different VMT values. Each of these clusters represents a group of scenarios from a single study, with each scenario in the cluster using the same growth assumptions but different land use and transportation elements.

Multivariate analysis

Testing the relative strength of all these variables simultaneously requires the application of a multivariate analytical method. Eighteen of the 2004–2010 scenario studies provided sufficient data to facilitate a multivariate analysis of 107 scenarios.

The dependent, or outcome, variable for the analysis is VMT, as measured by the percentage change in VMT between base and horizon years for each of the scenarios (see Figure 4.16). The mean percentage growth of VMT, 76 percent, may seem like a big increment of VMT growth, but it is over a mean forecast period of 31 years. The percentage change in VMT ranges from –4 percent, for a compact growth scenario in slow-growing Lansing, MI, to +348 percent for the trend scenario in fast-growing St. George, UT.

The independent, or explanatory, variables are divided into two sets: those that exhibit variance between scenarios and those that are constant between scenarios within the same study, but vary between studies. This is because scenarios that all come from the same region are not truly independent of

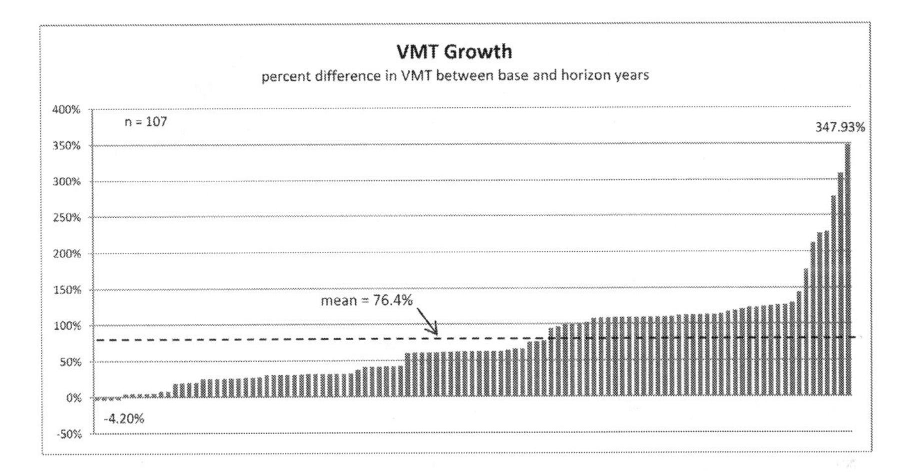

Figure 4.16 Percent Difference in VMT between Base and Planning Horizon Years.

each other, but will share many economic, geographic, and other characteristics that are not shared between scenarios from different regions.

The following variables are of the first type – i.e., they differ scenario to scenario, even within the same study:

- DENSITY is a continuous variable that measures the percentage change in density between base and horizon years.
- MIX is a dichotomous (a.k.a. dummy) variable that indicates whether or not a scenario emphasizes land use mixing.
- INFILL, also a dummy variable, designates whether a scenario focuses growth into central areas. INFILL is intended to represent the influence of destination accessibility.
- NODE is a dummy variable indicating whether a scenario focuses growth into activity centers or nodes. It, too, is intended to reflect destination accessibility.
- LANEMI is a continuous variable, measuring the percentage change between base and horizon years in lane miles of nonlocal roads.
- PRICING is a dummy variable, signifying whether a scenario incorporates transportation pricing policies.

The variables of the second type – that are common to all scenarios within a single study, but differ across studies – include the following:

- GROWTH is a continuous variable, measuring the percentage growth in population plus employment between base and horizon years.
- MODEL is an ordinal variable representing the degree to which the model utilized in a scenario project incorporates components that facilitate assessment of land use characteristics.

Because of this bi-level nature of the variables, we used a hierarchical model for the analysis. Within hierarchical models, each level in the data structure is represented by its own sub-model. Each sub-model captures the structural relations occurring at that level and the residual variability at that level. Sub-models at the different levels are linked statistically. A hierarchical model is, thus, able to account for the interdependence of scenarios in the same region and produce more accurate regression coefficient and standard error estimates (Bartholomew & Ewing, 2009; Raudenbush & Byrk, 2002).

In this analysis, the level 1 model relates VMT growth to scenario-specific characteristics plus a random error term. Thus, each region has a regression equation that describes the association between scenario characteristics and VMT growth within that region. The level 2 model treats the intercept and coefficients from level 1 as outcomes, and models them in terms of region-specific characteristics plus random effects.

Initially, all models assumed a "random intercept" form. Only the intercept term in the region-specific model was allowed to vary; all region-specific coefficients were taken as fixed. Then, this assumption was relaxed, and regression coefficients were allowed to vary as a function of region-specific characteristics. Equivalently, we can say that regional characteristics were allowed to interact with the scenario characteristics. This is termed a "random coefficients" model. Interactions between regional and scenario characteristics were seldom significant, with an exception noted below. Hence, the only results reported are for random intercept models.

The best-fit model is presented in Table 4.2. The model was estimated with no constant term (as a regression through the origin). If nothing changes from the base year, there should be no change in regional VMT.

Table 4.2 Hierarchical Model Predicting Percent Difference in VMT for 107 Scenarios

	Coefficient	*t*	*P*
Variables That Differ Scenario to Scenario			
DENSITY	−0.3	−4.26	< 0.001
MIX	−1.2	−0.36	0.72
INFILL	−1.44	−0.47	0.64
NODE	0.05	0.02	0.99
LANEMI	0.53	4.14	< 0.001
PRICING	−8.05	−2	0.048
Variables That Differ Project to Project			
GROWTH	1.08	2.12	0.049
MODEL	−0.19	−0.02	0.98

Because the continuous variables are percentage changes, their coefficients can be interpreted as elasticities. An elasticity is a percentage change in the dependent variable (VMT) with respect to a 1 percent increase in an independent variable (e.g., density).

Four coefficients are significantly different from zero, those of GROWTH, DENSITY, LANEMI, and PRICING. Each percent increase in population + employment is associated with a 1.08 percent increase in VMT. That is to say, VMT is slightly elastic with respect to population + employment growth. Each percent increase in density is associated with a 0.30 reduction in VMT. VMT is inelastic with respect to density, although not as inelastic as suggested by disaggregate travel studies (Ewing & Cervero, 2010). It seems likely that the density variable is soaking up effects of other variables that go hand-in-hand with density, particularly destination accessibility (Ewing & Cervero, 2010). Each percent increase in road lane miles is associated with a 0.53 percent increase in VMT. This elasticity falls somewhere between the short-run and long-run elasticities of VMT with respect to highway capacity estimated elsewhere in the literature (Cervero, 2002). All else being equal, the imposition of additional automobile user charges reduces VMT growth by 8.05 percent, a statistically significant reduction.

The other dummy variables are not significant, but with one exception, have the expected signs. Infill development is seen to reduce VMT growth by 1.44 percent. Mixed-use development reduces VMT growth by 1.20 percent. Nodal development is seen to increase VMT growth by 0.05 percent, the weakest of all the relationships and, counterintuitively, a positive relationship. It is not surprising that these dummy variables are not significant, as they crudely represent the underlying constructs and, being dichotomous variables, incorporate minimal variance.

The model sophistication variable, MODEL, also proves insignificant in most of the formulations tested. It clearly has no effect on the value of the intercept in the level 1 regression equation. That is to say, the extent to which a modeling system has been refined neither increases nor decreases forecasts of VMT growth. The model sophistication variable does interact with two of the level 1 variables, LANEMI and MIX, but with signs that are difficult to interpret, negative and positive, respectively. We would have expected that more sophisticated models would pick up effects of highway-induced traffic that increase VMT, and effects of mixed land uses that reduce VMT. We attribute the counterintuitive results to the imprecise nature of travel demand model characterizations.

We can estimate the effect of a shift to compact development on the growth of VMT by plugging realistic numbers into the model results in Table 4.2. If such a shift increases average regional density by 50 percent in 2050, emphasizes infill, mixes land uses to a high degree, and increases the price of automobile use, the model predicts that VMT will be 25 percent lower than projected under trend conditions ((–0.30 x 50) – (1.44 x 1) – (1.20 x 1) – (8.05 x 1)).

Twenty-five percent is very likely a conservative estimate for two reasons. First, limitations in the models and methods used to generate the data for this meta-analysis likely underestimated the degree to which the land use strategies in many of the scenarios would affect travel. Had we had more discriminating variables, rather than dummies, we might have seen larger effects. Second, all of the scenarios assumed the continuation of national and global economic and environmental trends, but it is very possible that these conditions will change in ways that would make continued reliance on personal vehicle travel less tenable, thereby increasing the difference between the alternative and trend scenarios. The impacts associated with climate change and peak oil, in particular, could have this effect. As noted in the New Visions 2030 scenario project from the Albany, New York, region (Capital District Transportation Committee, 2007, p. 44):

> If oil and gas remain widely available and relatively inexpensive, this would also support the likelihood of [trend] scenarios. However, if oil becomes scarce, and its price subsequently skyrockets, then we will have no choice but to significantly alter the manner in which we build and travel. Non-motor travel, such as walking and biking, will become more common. We will need to live close to where we work, while the kind of work we do will likely change dramatically. We will need to assemble our entire built environment much closer together, at higher densities, to try and eliminate long distance travel for everyday tasks. We will also be forced to localize our economy, including producing much of our food from within the local region. Under these conditions, [compact growth scenarios] would likely be closest to representing the kind of land development pattern that would result.

Other findings

One of the reasons articulated by the sponsors of land use-transportation scenario projects for focusing on more compact development is the anticipated savings in infrastructure costs associated with density. Consistent with the "cost of sprawl" literature (e.g., Burchell et al., 2002), the findings from the eight studies summarized in Figure 4.17 show substantial cost savings for virtually all of the denser alternative scenarios compared to their respective trend scenarios (N.B.: the one Salt Lake City scenario with greater costs is substantially less dense than the trend).

Another primary motivation articulated by scenario project sponsors is reducing the amount of agricultural land lost to suburban development (Figure 4.18). This rationale is particularly prevalent in regions facing possible future land scarcities.

Although VMT is the most often reported transportation statistic, many scenario projects also include measures of vehicle hours of travel (VHT,

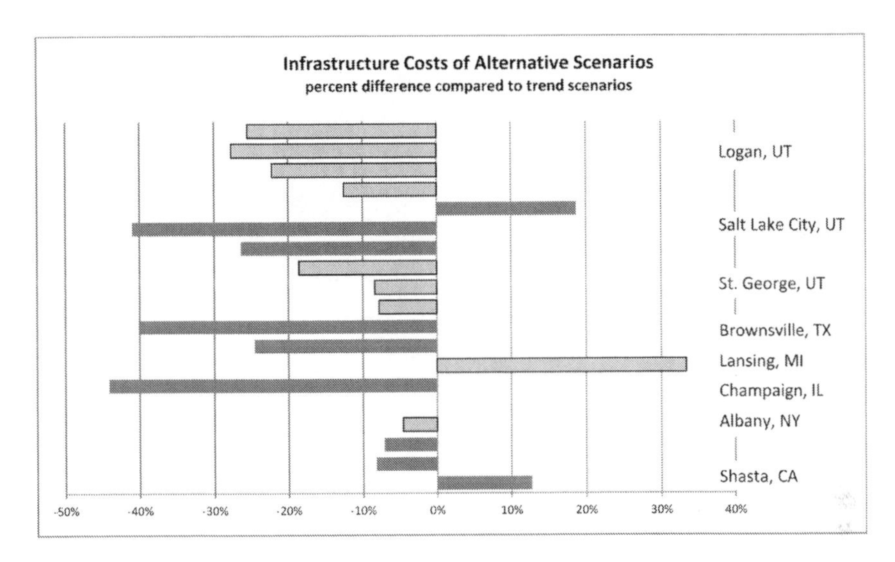

Figure 4.17 Percent Difference in Infrastructure Costs between Alternative and Trend Scenarios.

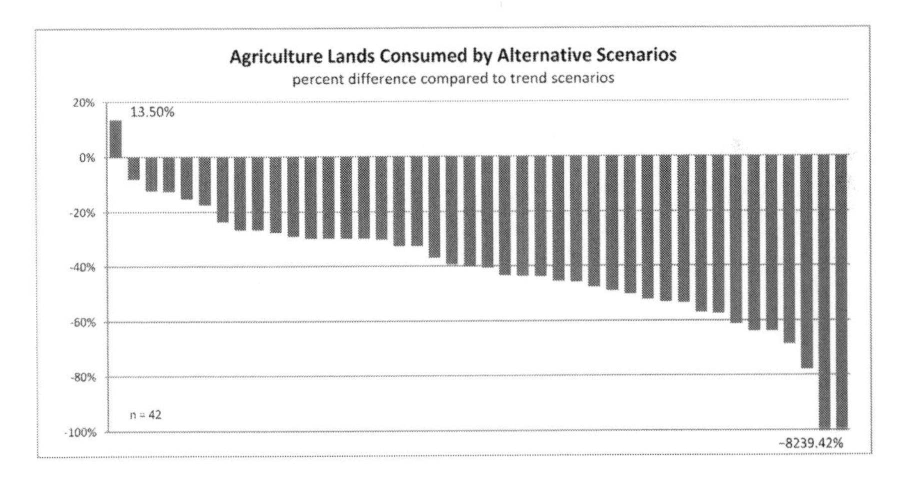

Figure 4.18 Percent Difference in Agricultural Land Consumption between Alternative and Trend Scenarios.

Figure 4.19) and vehicle hours of delay (VHD, Figure 4.20). Immediately noticeable is how similar the VHT chart is to the one for VMT (Figure 4.10), but how different the chart for VHD is from both the VMT and VHT charts. The implication is that alternative scenarios, which tend to be denser, are associated with higher levels of congestion.

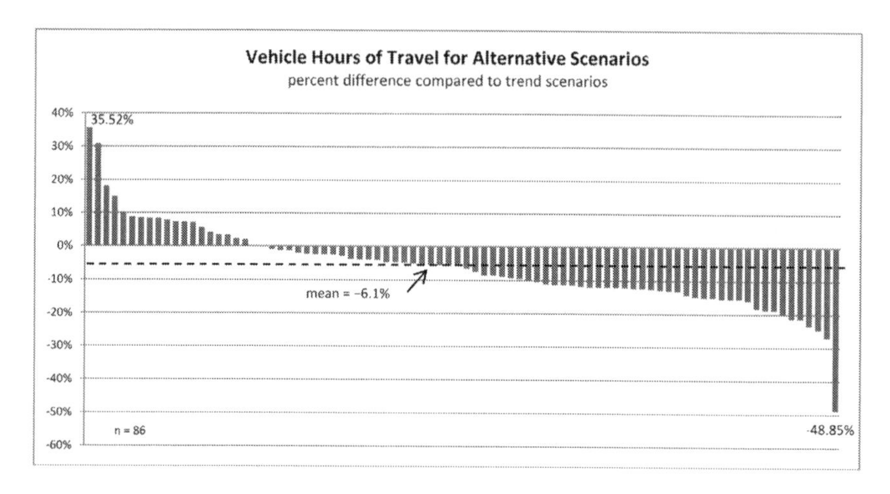

Figure 4.19 Percent Difference in Vehicle Hours of Travel: Alternative vs. Trend Scenarios.

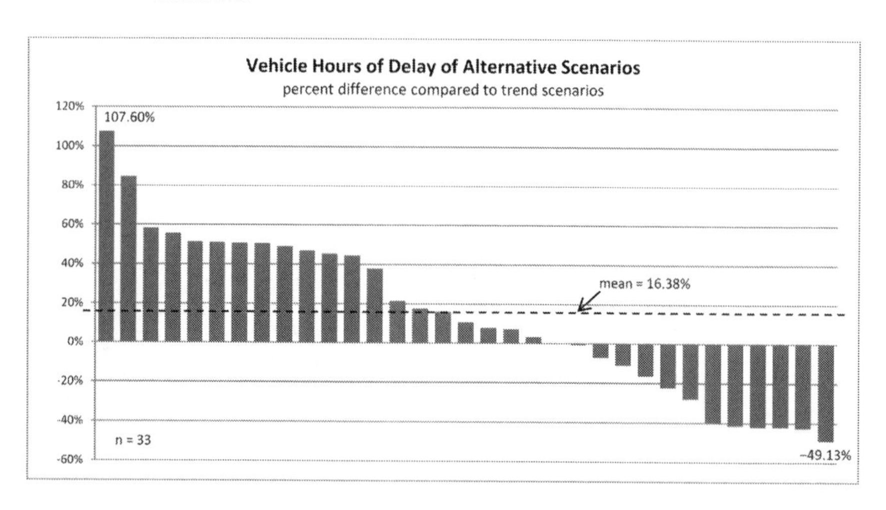

Figure 4.20 Percent Difference in Vehicle Hours of Delay: Alternative vs. Trend Scenarios.

Incorporating scenario outputs into transportation planning processes

Planning for metropolitan-wide transportation systems in the United States was institutionalized by the 1962 Federal Aid Highway Act. The planning practice that grew out of that Act, which generally did not include land use-transportation scenario techniques, remained largely unchanged until the passage of the Intermodal Surface Transportation Efficiency Act ("ISTEA")

30 years later. ISTEA revolutionized systems planning, principally, by allowing for greater flexibility in how federal transportation funds could be used, and by requiring MPOs to incorporate a more expansive list of planning factors into the development of systems plans. Included in this list was consideration of (1) the possible effects of transportation investments on development patterns and (2) the consistency of transportation plans with land use and development plans. Many of ISTEA's innovations were carried forward, first into the Transportation Equity Act for the 21st Century ("TEA-21"), then into the Safe, Accountable, Flexible, Efficient Transportation Equity Act: A Legacy for Users ("SAFETEA-LU") and the Moving Ahead for Progress in the 21st Century Act ("MAP-21"). MAP-21 included voluntary provisions related to scenario planning. Pursuant to Section 450.324 of the MAP-21 regulations, MPOs may use scenario planning to inform decision makers about the implications of various investments and policies on transportation system performance. As the preamble to the MAP-21 regulations notes, "Scenario planning is currently used by many MPOs as part of their transportation planning process, and FHWA and FTA consider it a best practice." These innovations have been continued in the most recent law, Fixing America's Surface Transportation (FAST) Act.

By directing MPOs to look at land use and transportation interactions, ISTEA encouraged some MPOs to engage in land use-transportation scenario planning. What was unclear from ISTEA was how the output of scenario analysis might feed into standard long-range planning processes. Federal regulations require MPOs to base updates to long-range plans on "the latest available estimates and assumptions for population, land use, travel, employment, congestion, and economic activity" (23 C.F.R. 450.322(e)). In the context of a scenario analysis, what does the phrase "latest available estimates and assumptions" mean? Does it require the use of trend land use assumptions, notwithstanding that the scenario process may have resulted in the rejection of the trend scenario and the selection of a smart growth–type alternative scenario?

There is evidence from scenario projects conducted in the 1990s that a number of MPOs thought this was the case. More recent projects, however, point to a different approach. In Sacramento, California, for example, the MPO concluded its Sacramento Region Blueprint scenario study by working individually with the region's local governments to gauge each jurisdiction's level of commitment to the study's preferred scenario. As governments endorsed the scenario, the MPO altered the assumptions about future growth for that jurisdiction to match those from the preferred scenario, rather than from the trend projection. Other projects from the 2004–2010 time period have replicated this method.

As important as these developments are, however, they reflect only a small minority of the nation's metropolitan areas. As in other policy areas, the real action in promoting scenario planning is happening at the state level. Justice William Brandeis famously wrote, "It is one of the happy incidents of the federal system that a single courageous state may, if its citizens choose, serve as a laboratory and try novel social and economics experiments without risk to the rest of the country" (New State Ice Co., 1932, p. 311). So it seems to be the case with climate policy and as states advance climate policy, they have also begun to advance scenario planning techniques.

More than any other state, California has taken a leading role in the development of climate change laws and policies. One of those laws, Senate Bill 375 (2008), was motivated by the state legislature's understanding that "without improved land use and transportation policy, California will not be able to achieve the goals of [its other climate change policies]" (SB 375 § 1(c)). The bill, which has been described both as a "bold experiment" (Nichols, 2009) and a "necessary collision" (Stern, 2008), requires the state to set greenhouse gas emission reduction targets for passenger vehicles for each of the state's 18 MPOs. The MPOs are then required to adopt Sustainable Communities Strategies (SCSs) that meet the targets. If a region does not or cannot meet the emission targets, the MPO is obliged to create an Alternative Planning Strategy (APS) "showing how those greenhouse gas emission targets would be achieved through alternative development patterns, infrastructure, or additional transportation measures or policies" (Cal. Gov't Code § 65080(b)(2)(H)). Although SB 375 does not specifically mandate the use of scenario planning, the method is implied by several sections of the bill.

Another state law requirement, this one coming from Oregon, goes even further in promoting scenario planning. HB 2001 (2009) directs Metro, the Portland-area MPO, to "develop two or more alternative land use and transportation scenarios that accommodate planned population and employment growth while achieving a reduction in greenhouse gas emissions from motor vehicles" (HB 2001 § 37(2)(a) (2009)). The emission reductions targets to be achieved by these scenarios must be based on the state's adopted goal of reducing emissions 10 percent below 1990 levels by 2020 and 75 percent below 1990 levels by 2050 (Ore. Rev. Stat. § 468A.205). Metro is required to formally adopt one of the land use-transportation scenarios and local governments in the metropolitan area must then amend their comprehensive plans and zoning ordinances to be consistent with the adopted scenario (HB 2001 § 37(3) (2009)). Senate Bill 1059, passed during a special interim session in 2010, creates a process for the state transportation and land use agencies to craft similar scenario planning mechanisms for the state's five other MPOs.

Conclusion

Land use-transportation scenario planning is an important planning tool whose popularity continues to increase. The technique's association with articulating more compact alternatives for future growth is further demonstrated by the projects summarized in this chapter. These projects show, with stronger statistical evidence than in previous studies (e.g., Bartholomew & Ewing, 2009), how compact growth alternatives can increase regional livability by reducing vehicle travel demand. The more recent projects also demonstrate how scenario techniques can be effectively integrated into traditional long-range regional transportation planning processes.

These important advances in regional scenario practice are offset, to some degree, by several areas where limitations remain. Transportation modeling capacity continues to be a concern. Whereas several MPOs mentioned in this chapter host state-of-the-art modeling systems, most have only limited capacity to assess the impacts of land use-transportation scenarios. Another area of concern is the failure to incorporate important changes in global economic and environmental conditions, such as climate change, both as input variables and as evaluation metrics. The current practice of land use-transportation scenario planning recognizes that single-allocation land use forecasts were based on a fictional assumption that land use patterns could be reliably predicted. The practice should now recognize that global economic and environmental conditions underlying planning analyses are similarly unpredictable.

5 Quantitative analysis of historical VMT growth

Key to understanding the emerging planning trends outlined in this book is the issue of transportation consumption. The amount that people travel today, and are expected to travel in the future, is a central factor affecting many of the sustainability and quality-of-life goals of public policy. For example, the national goals of the federal aid highway system include:

Safety – To achieve a significant reduction in traffic fatalities and serious injuries on all public roads.
Infrastructure condition – To maintain the highway infrastructure asset system in a state of good repair.
Congestion reduction – To achieve a significant reduction in congestion on the National Highway System.
Environmental sustainability – To enhance the performance of the transportation system while protecting and enhancing the natural environment.

All of these goals relate, either directly or indirectly, to the amount of vehicle travel occurring on the nation's roads – safety, congestion, fuel consumption, and air pollution are all affected significantly by the amount that people drive. Measuring transportation consumption is most frequently accomplished by an assessment of vehicle mile traveled or VMT – the sum of all miles traveled by all vehicles over a specified time period and geographic area.

This chapter expands on previous work, using more recent data, additional metrics, and structural equation modeling to explain VMT levels of urbanized areas and to test the effects of various policy and planning levers. It concludes with best-estimate elasticities of VMT per capita with respect to these levers.

Literature review

There are rich literatures relating VMT to land use, highway capacity, the real price of fuel, and transit access. The literatures on the first three topics are so extensive we will limit this review to meta-analyses. Unlike traditional research methods, meta-analyses use summary statistics from individual primary studies as the data points in a new analysis.

Built environment and VMT

In travel research, urban development patterns have come to be characterized by "D" variables. The original "three Ds," coined by Cervero and Kockelman (1997), are density, diversity, and design. The Ds have multiplied since Cervero and Kockelman's original article, with the addition of destination accessibility and distance to transit (Ewing & Cervero, 2001; Ewing & Cervero, 2010). Whereas not part of the environment, demographics are another D in travel studies, controlled as confounding influences.

A meta-analysis uncovered more than 200 studies of the built environment and travel (Ewing & Cervero, 2010). Of these, 60 studies yielded usable outcome measures from which to compute weighted average elasticities. An elasticity is a measure of effect size equal to the percentage change in an outcome variable (such as VMT) with respect to a 1 percent increase in an explanatory variable (such as density). The D variable that is most strongly associated with VMT is destination accessibility. In fact, the -0.19 VMT elasticity is nearly as large as the elasticities of the first three D variables – density, diversity, and design – combined.

Next-most strongly associated with VMT are design metrics expressed in terms of intersection density or street connectivity. The elasticities of these two street network variables are fairly similar. Both short blocks and frequent intersections shorten travel distances, apparently to about the same extent. Surprisingly, population density was found to be weakly associated with travel behavior once these other variables are controlled. In an effort to explain the much higher elasticities reported in the literature, Ewing and Cervero (2010) note: "The relatively weak relationships between density and travel likely indicate that density is an intermediate variable that is often expressed by the other Ds (i.e., dense settings commonly have mixed uses, short blocks, and central locations, all of which shorten trips and encourage walking)" (p. 276).

Highway capacity and VMT

Based on a meta-analysis of the VMT-inducing effects of highway expansion, Cervero (2002, p. 17) concludes that "the preponderance of research suggests that induced-demand effects are significant, with an appreciable share of added capacity being absorbed by increases in traffic, with a few notable exceptions."

In the short run a variety of sources contribute to increased traffic without any highway-induced development. These include changes in route, mode, time of travel, and destination. In addition, there is the possibility of new trips that would not have occurred without the new infrastructure capacity.

In the long run, increases in highway capacity may improve accessibility to developable lands and lower travel times to the point where residences and businesses are drawn to locate near the expanded highway capacity (Ewing, 2008). Cervero (2002) computes a long-run elasticity of VMT with respect to highway capacity of between 0.63 and 0.73.

Fuel prices and VMT

The meta-analytical literature on VMT growth with respect to the real price of fuel is sparse. The primary work in the area is Graham and Glaister's (2004) review of more than 50 studies measuring the fuel price elasticities for car trips and car kilometers within European Union countries. Looking at both short-term (less than 1 year) and long-term effects, the researchers found that the unweighted mean short-run elasticities for trips and kilometers across the studies were roughly equivalent at –0.16. Over time, however, the two measures diverged, with trips decreasing only slightly to –0.19, but kilometers dipping substantially to –0.31. A parallel study by Goodwin, Dargey, and Hanly (2004) summarizing 69 studies from Europe and North America came to similar conclusions, with a mean short-term vehicle-km elasticity of –0.1 and a long-term elasticity of –0.29.

Meta studies of gasoline demand versus price are more numerous, and given that gasoline demand is a rough proxy for VMT, particularly in the short run, this literature sheds light on the fuel price–VMT relationship. One meta-analytic study derived a long-run mean price elasticity of gasoline demand of –0.53 (Brons, Nijkamp, Pels, & Rietveld, 2006). Another meta-analysis of gasoline price elasticities based on hundreds of studies across the globe found a mean short-run elasticity of –0.23 and a mean long-run elasticity of –0.58 (Espey, 1998). This study concludes with this relevant thought: "The finding of different elasticity estimates using data prior to 1974 and data after 1974 suggests the need for updated studies and for care to be taken in extrapolating into the future using elasticity estimates from the 1970s or even the 1980s" (p. 294).

In an oft-cited more recent study, which overcomes some of the methodological limitations of earlier studies, Small and Van Dender (2007) observed a low (under –0.10) short-run price elasticity of gasoline demand. But importantly, they found gasoline's long-run price elasticity to be much higher, approximately –0.43. Also, they found that the elasticity of VMT with respect to fuel cost per mile (controlling for increased vehicle fuel efficiency) was roughly half the price elasticity of gasoline demand.

Transit service and VMT

Historically, research examining the role of public transit in reducing VMT and greenhouse gas (GHG) emissions has focused directly on mode shifts from driving to transit occurring as a result of transit investments. Such research typically shows only modest reductions in vehicle travel. However, a growing body of research suggests that cities with comprehensive transit facilities achieve more efficient use of their transportation systems that is not fully captured by mode shifts from driving to transit. This concept, commonly referred to as transit leverage or the land use multiplier effect, states that one-mile traveled on transit corresponds to a disproportionately higher reduction in automobile travel. The multiplier is typically expressed as VMT reduced per passenger-mile of transit or as a multiplier of the mode shift effects of transit.

In other words, the influences of transit – including more compact and mixed land uses in station areas, a higher propensity by users to chain trips, reduced traffic congestion, and a significantly higher rate of related nonmotorized travel (walk and bike trips) – converge to reduce automobile travel and GHG emissions to a greater degree than simply the distance traveled via transit. Even those who live near transit but do not utilize it may drive less due to the compact, mixed-use neighborhoods and opportunities to walk and bike fostered by transit. Figure 5.1 illustrates how the land use multiplier relates to the other ways in which transit produces and displaces GHG emissions.

The mechanism by which transit leverages larger reductions in VMT is straightforward. Transit creates opportunities for transit-oriented development (TOD), "compact, mixed-use development near transit facilities with

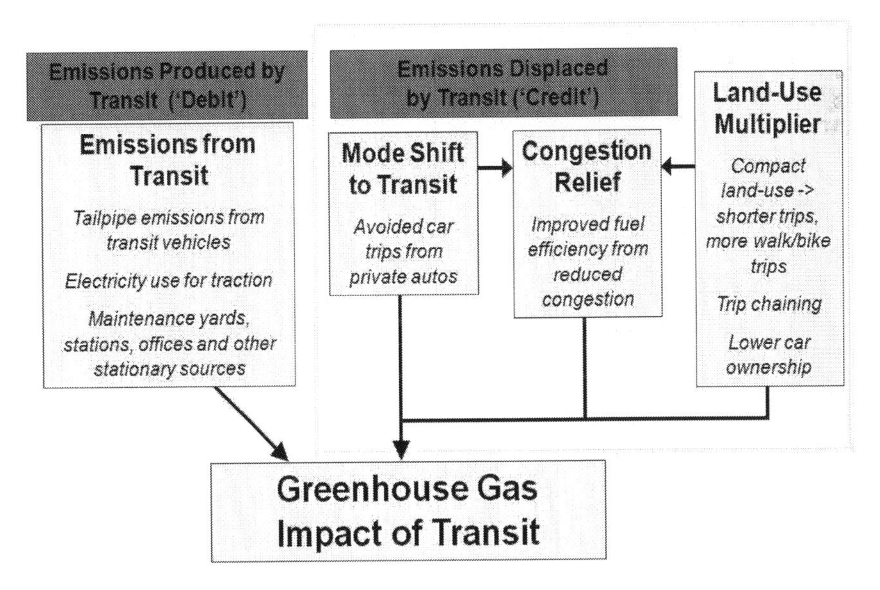

Figure 5.1 Overview of APTA Approach to Estimating the GHG Impacts of Public Transit.

(APTA, 2009, p. 8).

high-quality walking environments" (Cervero et al., 2004, p. S-1), which by definition combines all of the D variables.

However, researchers have yet to reach a consensus on the magnitude of the land use multiplier effect. Studies have produced estimates for the land use multiplier ranging from 1.29 to 9 (Ewing and Hamidi, 2014). Estimates of the land use multiplier can even vary widely within a given study.

Parallel analyses

The book *Growing Cooler* (Ewing, Bartholomew, Winkelman, Walters, & Chen, 2008) asked and attempted to answer the question: How does compact development affect VMT and associated greenhouse gas emissions that contribute to global warming? Using structural equation modeling and both cross-sectional and longitudinal data for 84 large U.S. urbanized areas, Chapter 8 estimated elasticities of VMT with respect to population, real per capita income, population density, highway lane miles, transit revenue miles, transit passenger miles, and the real price of fuel (see Table 5.1). Table 5.1 suggests, for example, that a 1 percent increase in highway lane miles will bring about a 0.55 percent increase in VMT.

More recently, Cervero and Murakami (2010) similarly used structural equation modeling, plus cross-sectional data from 370 U.S. urbanized areas, to estimate elasticities of VMT per capita with respect to household income, population density, road density, rail density, and other land use variables related to density and accessibility. Their results are presented in Table 5.2. They are generally consistent with the results of Ewing et al. (2008), although the elasticity of roadway density is smaller and the elasticity of population density is larger.

Table 5.1 Elasticities of VMT with Respect to Urban Variables

	Cross-Sectional Analysis	*Longitudinal Analysis*	*Best Estimate*
Population	0.97	0.874	0.95
Real per capita income	0.531	0.538	0.54
Population density	−0.213	−0.152	−0.30
Highway lane miles	0.463	0.684	0.55
Transit revenue miles	−0.075	−0.023	−0.06
Transit passenger miles	−0.068	−0.03	−0.06
Heavy-rail miles	−0.013	−0.021	−0.01
Light-rail miles	−0.003	−0.002	NA
Real fuel price	NA	−0.171	−0.17

Source: Adapted from Ewing et al. (2008, p. 123).

Table 5.2 Elasticities of VMT per Capita with Respect to Urban Variables

	Estimate
Household income	0.21
Population density	-0.38
Roadway density	0.42
Rail density	-0.003
Urbanized area	0.02
Percent commuting by auto	0.6

Source: Adapted from Cervero andMurakami (2010, p. 411).

Update and expansion

This chapter updates and expands the Ewing et al. (2008) and Cervero and Murakami (2010) analyses. It updates in the sense that relationships are estimated through 2010, whereas the earlier analyses ran only through 2005 and 2003, respectively.

It expands in two ways. First, this analysis distinguishes between freeways and other main highways and streets on the assumption that the two types of roadway capacity may have different effects on VMT. Whereas freeway capacity may increase VMT by inducing traffic and sprawl, arterial and collector mileage may have less induced effect and may allow more direct routing of traffic in a more complete grid. It also distinguishes between heavy rail and light rail mileage, which could have different effects on the built environment and VMT. Also, the new analysis replaces a single transit service measure, transit revenue miles per capita, with two measures, one representing service coverage and the other service frequency. Average service coverage is roughly measured in terms of route miles of service divided by urbanized area in square miles. Average service frequency is roughly measured in terms of revenue miles of service divided by route miles of service. These are distinct service dimensions, essentially uncorrelated.

Second, the new analysis expands the sample of urbanized areas, from 84 urbanized areas for Ewing et al. (2008) and 370 urbanized areas for Cervero and Murakami (2010) to 443 areas in this study. The initial sample includes all of the urbanized areas in the United States. Some were lost to the sample for lack of complete data sets, for lack of transit service, or for lack of freeway capacity. The final sample of 315 urbanized areas represents 82 percent of the nation's urban population and 65 percent of the nation's total population.

This analysis differs from the Ewing et al.'s analysis in other respects. In *Growing Cooler*, VMT was measured as the sum of VMT on freeways plus VMT on arterials, as estimated by the Texas Transportation Institute (TTI). In this analysis, VMT is measured as the total for all classes of roadways in urbanized areas, as reported in the FHWA's *Highway Statistics*. It is a broader measure of VMT, and when compared to TTI's estimates for the same period, is plausibly larger for nearly all urbanized areas, as it should be.

Methodology

Research design

In this study, a cross-sectional model is estimated to capture the long-run relationships between transportation and land use at a point in time, 2010. Each urbanized area has had decades to arrive at quasi-equilibrium among land use patterns, road capacity, transit capacity, and VMT. This quasi-equilibrium is captured via structural equation modeling (SEM).

Method of analysis

SEM is a statistical technique for evaluating complex hypotheses involving multiple, interacting variables (Grace, 2006). The estimation of SEM models involves solving a set of equations. There is an equation for each "response" or "endogenous" variable in the system. They are affected by other variables, and may also affect other variables. Variables that are solely predictors of other variables are termed "influences" or "exogenous" variables. They may be correlated with one another but are determined outside the system.

Typically, solution procedures for SEM models focus on observed versus model-implied correlations in the data. The unstandardized correlations or co-variances are the raw material for the analyses. Models are automatically compared to a "saturated" model (one that allows all variables to intercorrelate), and this comparison allows the analysis to discover missing pathways and, thereby, reject inconsistent models.

Data

We gathered data from several primary sources for our cross-sectional analysis. For the sake of consistency, the boundaries used to compute explanatory variables had to be the same as the boundaries used to estimate our dependent variable, VMT per capita, from FHWA *Highway Statistics*.

The *Highway Statistics* definition of urbanized area is different than the census definition. According to FHWA, "the boundaries of the area shall encompass the entire urbanized area as designated by the U.S. Bureau of the Census plus that adjacent geographical area as agreed upon by local officials in cooperation with the State" (Federal Highway Administration 2003, p. 6). Cervero and Murakami (2010) used the census boundaries for their analysis and deleted urbanized areas from the sample if the census and FHWA boundaries were hugely different. We chose not to make such approximations or lose many cases, and therefore set out to find FHWA adjusted boundaries for urbanized areas in a geospatial shapefile format, which we could then use to conduct spatial analyses in GIS (see Figure 5.2).

FHWA advised us to contact individual state DOT offices for their shapefiles, which we did. This sometimes required several calls to find the right office. In this way, we were able to obtain shapefiles for all 50 states and 443 urbanized areas. We then combined the individual state files into one national shapefile by using the "merge" function in GIS. Many of the urbanized areas

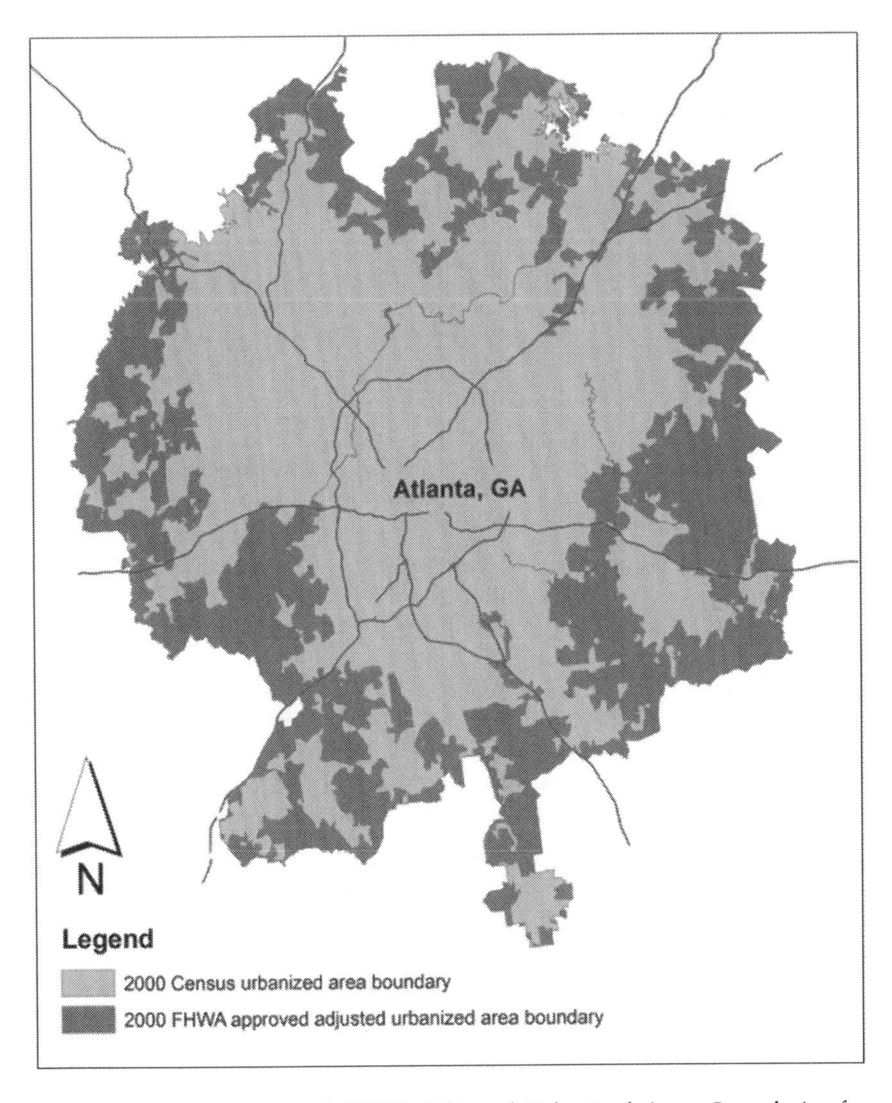

Figure 5.2 2000 Census and FHWA-Adjusted Urbanized Areas Boundaries for Atlanta.

(Ewing et al., 2014, p. 3086).

cross state boundaries and in this case we had more than one polygon for each urbanized area. So, we used the "dissolve" function in GIS to integrate those polygons into one for each urbanized area.

After cleaning the data, we did several spatial joins in GIS to capture data from other sources. For example, we used the "centroid" function to join 2010 census tracts to FHWA-adjusted urbanized areas. We then aggregated values of per capita income for census tracts to obtain urbanized area weighted averages (weighted by population).

Variables

The variables in our models are defined in Table 5.3. The variables fall into three general classes:

Our outcome variable, VMT per capitaExogenous explanatory variables. The exogenous variables, population and per capita income, are determined by regional competitiveness. The real fuel price is determined by federal and state tax policies and regional location relative to ports of entry and refining capacity. Variables representing highway capacity and rail system capacity were also treated as exogenous, as they are the result of long-lived policy decisions to invest in highways or transit.

Endogenous explanatory variables. The endogenous variables are a function of exogenous variables and are, in addition, related to one another. They depend on real estate market forces and regional and policy decisions: whether to increase transit revenue service, whether to zone for higher densities.

All variables were transformed by taking natural logarithms. The use of logarithms has two advantages. First, it makes relationships among our variables more nearly linear and reduces the influence of outliers (such as New York and Los Angeles). Second, it allows us to interpret parameter estimates as elasticities, which summarize relationships in an understandable and transferable form.

Model

Our SEM model was estimated with the software package Amos (version 7.0, SPSS 2007) and maximum likelihood procedures. The path diagram in Figure 5.3 is copied directly from Amos. Causal pathways are represented by uni-directional straight arrows. Correlations are represented by curved bi-directional arrows (to simplify the already complex causal diagram, some correlations are omitted). By convention, circles represent error terms in the model, of which there is one for each endogenous (response) variable.

Most of the causal paths shown in the path diagram are statistically significant (have nonzero values). The exceptions are a few paths that are theoretically significant, although not statistically significant.

Table 5.3 Variables Included in the Urbanized Area Model

Variable	*Definition*	*Source*	*Mean*	*Standard Deviation*
Dependent Variable				
vmt	Natural log of daily VMT per capita	FHWA *Highway Statistics*	3.09	0.25
Exogenous Variables				
pop	Natural log of population (in thousands)	U.S. Census	12.45	1.16
inc	Natural log of income per capita	American Community Survey	10.13	0.19
fuel	Natural log of average fuel price metropolitan average fuel price	Oil Price Information Service	1.03	0.06
flm	Natural log of freeway lane miles per 1,000 population	FHWA *Highway Statistics*	–0.46	0.53
olm	Natural log of other lane miles per 1,000 population	FHWA *Highway Statistics* NAVTEQ	0.91	0.32
hrt	Directional route miles of heavy-rail lines per 100,000 population*	National Transit Database	0.04	0.23
lrt	Directional route miles of light-rail lines per 100,000 population*	National Transit Database	0.09	0.33
Endogenous Variables				
popden	Natural log of gross population density	U.S. Census	7.33	0.44
rtden	Natural log of transit route density per square mile	National Transit Database	0.67	0.82
tfreq	Natural log of transit service frequency	National Transit Database	8.51	0.59
tpm	Natural log of annual transit passenger miles per capita	National Transit Database	3.76	1.12

* 1 was added to values so that urbanised areas with no rail mileage would have a zero value when log transformed.

The main goodness-of-fit measure used to choose among models was the chi-square statistic. Probability statements about an SEM model are reversed from those associated with null hypotheses. Probability values (*p*-values) used in statistics are measures of the degree to which the data are unexpected, given the hypothesis being tested. In null hypothesis testing, a finding of a p-value < 0.05 indicates that we can reject the null hypothesis because the data are very unlikely to come from a random process. In SEM, we seek a model with a small chi-square and large p-value (> 0.05) because that indicates that the data are not unlikely given that model (i.e., the data are consistent with the model).

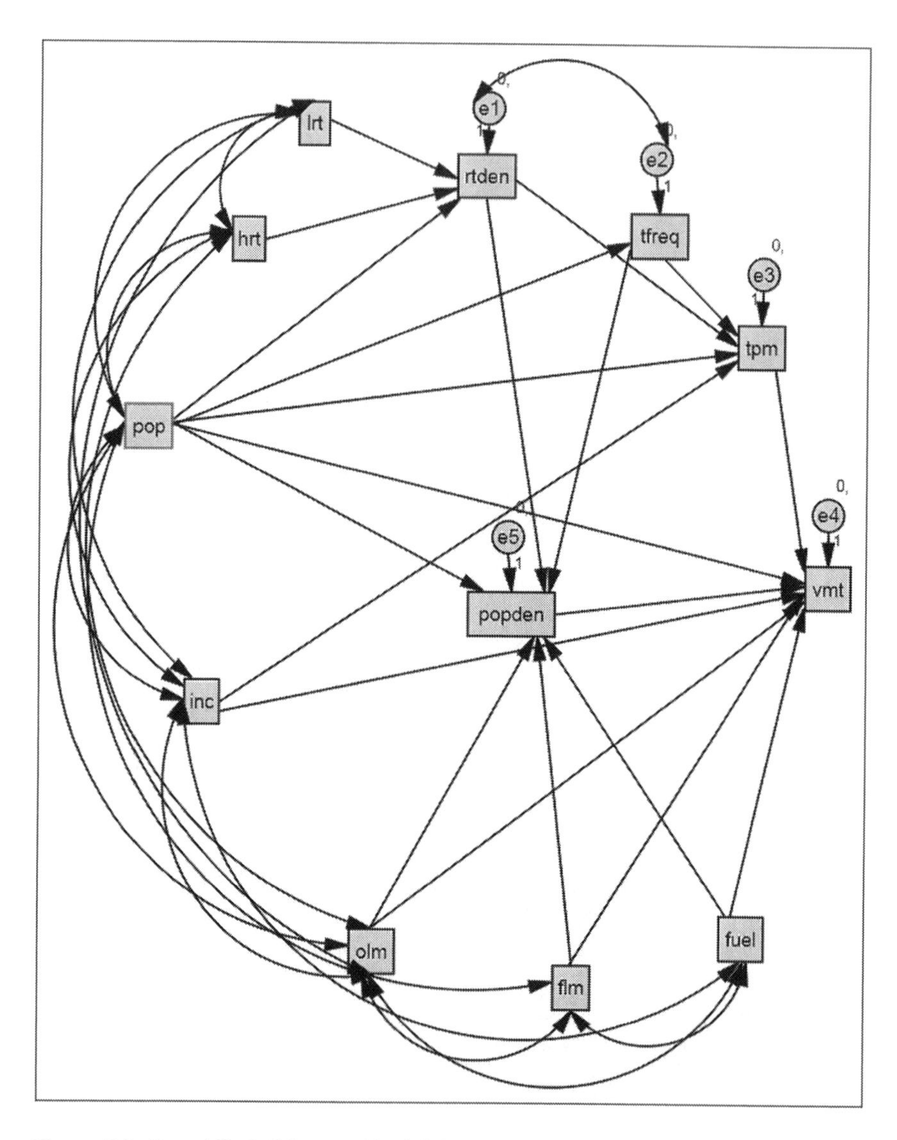

Figure 5.3 Causal Path Diagram Explaining VMT per capita for Urbanized Areas. (Ewing et al., 2014, p. 3089).

Results

The VMT model in Figure 5.3 has a chi-square of 26.5 with 22 model degrees of freedom and a p-value of 0.23. The low chi-square relative to model degrees of freedom and a high (> 0.05) p-value are indicators of good model fit.

The regression coefficients in Table 5.4 give the predicted effects of individual variables, all other things equal. These are the direct effects of one variable on another. They do not account for the indirect effects through other endogenous variables. Also of interest are the total effects of different variables on VMT per capita, accounting for both direct and indirect pathways (see Table 5.5).

Population growth is a driver of VMT growth. As urbanized areas grow, destinations tend to become farther apart (e.g., the suburbs are farther from the CBD). Therefore, the direct effect of population size on VMT per capita is positive and significant due to the simple fact of their size. At the same time, as urbanized areas grow, they become denser and shift away from a singular focus on road capacity to meet travel demands toward a balance of roads and transit. The indirect effect of population size on VMT per capita is negative.

Another exogenous driver of VMT growth is income. As per capita income rises, people travel more by private vehicle, reflecting the general wealth of

Table 5.4 Path Coefficient Estimates (Regression Coefficients) and Associated Statistics for Direct Effects in the 2010 VMT per capita Model (see Figure 5.3)

			Coeff	*S.E.*	*C.R.*	*P*
tfreq	< –	pop	0.235	0.025	9.234	< 0.001
rtden	< –	lrt	0.495	0.131	3.787	< 0.001
rtden	< –	hrt	0.355	0.187	1.9	0.057
rtden	< –	pop	−0.103	0.042	−2.463	0.014
popden	< –	olm	−0.552	0.047	−11.748	< 0.001
popden	< –	rtden	0.197	0.017	11.528	< 0.001
tpm	< –	pop	0.141	0.041	3.44	< 0.001
tpm	< –	tfreq	0.796	0.077	10.406	< 0.001
popden	< –	tfreq	0.187	0.023	8.035	< 0.001
tpm	< –	rtden	0.839	0.049	17.124	< 0.001
popden	< –	flm	−0.108	0.02	−5.383	< 0.001
tpm	< –	inc	0.902	0.208	4.345	< 0.001
popden	< –	pop	0.066	0.011	5.849	< 0.001
popden	< –	fuel	0.733	0.236	3.111	0.002
vmt	< –	fuel	−0.448	0.238	1.883	0.06
vmt	< –	popden	−0.238	0.043	−5.577	< 0.001
vmt	< –	olm	0.04	0.051	0.784	0.433
vmt	< –	flm	0.133	0.021	6.412	< 0.001
vmt	< –	inc	0.304	0.062	4.889	< 0.001
vmt	< –	tpm	−0.016	0.011	−1.427	0.154
vmt	< –	pop	0.078	0.012	6.635	< 0.001

Table 5.5 Direct, Indirect, and Total Effects of Variables on VMT per capita in the Cross-Sectional Model for 2010 (see Figure 5.3)

	Direct	*Indirect*	*Total*
pop	0.078	–0.025	0.052
inc	0.304	–0.015	0.289
fuel	–0.448	–0.175	–0.623
hrt	0	–0.021	–0.021
lrt	0	–0.03	–0.03
flm	0.133	0.026	0.159
olm	0.04	0.131	0.172
popden	–0.238	0	–0.238
rtden	0	–0.06	–0.06
tfreq	0	–0.057	–0.057
tpm	–0.016	0	–0.016

the community. The direct effect of per capita income on VMT per capita is positive and highly significant. Income has an indirect effect as well, through transit passenger miles per capita. Surprisingly, the effect of income on transit use is positive, hence the indirect effect on VMT is negative. Wealthier communities may provide more transit service, and higher-income residents in large regions such as New York may use transit to commute in from the suburbs.

Controlling for other influences, areas with more freeway capacity are significantly less dense and have significantly higher VMT per capita. Areas with more highway capacity in arterials, collectors, and local streets are also significantly less dense (which affects VMT per capita indirectly) but the direct effect of other highway capacity on VMT per capita is not significant. From the standpoint of induced traffic, other roadways are more benign than freeways.

Transit has an effect opposite to that of highways. Areas with more service coverage and more service frequency have higher development densities, which leads to lower VMT per capita. They also have more transit passenger miles per capita, which leads to lower VMT per capita. The causal path through transit passenger miles constitutes the direct effect of transit on VMT. The causal path through development density constitutes the indirect or land use effect of transit on VMT. The latter divided by the former is the land use multiplier.

The two rail variables, HRT and LRT directional route miles per capita, are positively associated with route coverage, and through that variable, increase transit passenger miles per capita and reduce VMT per capita. Surprisingly, neither HRT route mileage nor LRT route mileage has a direct effect on the development density of urbanized areas. One possible explanation for the

Table 5.6 Transit's Effect on VMT

Service Coverage (rtden)	
Indirect effect	–0.05
Direct effect	–0.01
Land use multiplier	3.49
Service Frequency (tfreq)	
Indirect effect	–0.04
Direct effect	–0.01
Land use multiplier	3.49

failure of rail to raise densities is the oft-cited potential of rail extensions into the suburbs to cause sprawl, as long-distance commuters park and then ride into the city.

The real fuel price is negatively associated with VMT per capita, both directly and indirectly through an effect on development densities. The direct price elasticity, around –0.45, is what one would expect from the literature (the long-run elasticity being much greater than the short-run elasticity). There are persistent regional variations in real fuel prices and these appear to affect both urban form and VMT per capita.

Urbanized area density is negatively related to VMT per capita. The elasticity, –0.24, suggests that every 1 percent rise in density is associated with a 0.24 percent decline in VMT per capita. With density serving as a proxy for all the D variables (density, diversity, design, and destination accessibility), the elasticity looks reasonable.

We are also interested in the size of transit's land use multiplier. The two transit variables, which as noted previously, are uncorrelated, have remarkably similar relationships to VMT. As shown in Table 5.6, the land use multiplier is 3.49 for both variables.

Discussion and conclusion

As debates about air quality, energy, and climate policy have heated up, increased attention has been paid to the roles of urban form and transit infrastructure in addressing these policy challenges. The vigor that has accompanied research in the area, however, has sometimes given rise to warnings against over-exuberance. Whereas acknowledging that land development patterns likely have an influence on travel, a special Transportation Research Board panel signaled that it did not have as much "verifiable scientific evidence" as it would have liked to support its conclusions (TRB, 2009, p. 131), conclusions that have been criticized by some as unnecessarily conservative (Ewing, Greenwald, & Zhang, 2011).

The analysis presented in this chapter does not, of course, address all of the cautions and criticisms. It does, however, advance the state of research in some significant ways. By using data from 315 different urbanized areas, the analysis provides a nationally comprehensive assessment, covering nearly two-thirds of the U.S. population. Moreover, rather than focusing on just one factor that affects travel demand, the analysis provides a holistic approach that integrates all of the major groups of influences: demographics, development patterns, system capacities, and transportation costs (see Bartholomew, 2009).

Naturally, the analysis has its limitations, suggesting several possible future investigations. Adding a longitudinal analysis is an obvious place to focus future research. This might be facilitated by the development of a richer and longer-term database. Another improvement would be the introduction of multilevel modeling capacities into the SEM framework to account for the fact that changes in conditions over time are occurring within each urban area, as well as between urban areas.

Limitations notwithstanding, the integrated approach used here has led to several important findings: freeway expansions seem to have stronger induced demand effects than arterial expansions; increases in development densities and fuel costs are, in fact, associated with reduced driving, and in some cases the association is stronger than previously measured. Transit service coverage and service frequency have direct and indirect effects on VMT, the latter much larger in magnitude than the former. These observations provide a platform for understanding of how different policy options might work on the ground.

6 Qualitative analysis of regional transportation plans

Introduction

In this chapter, we conclude our analysis of metropolitan transportation planning practice by reviewing the qualitative results of our analysis of region transportation plans (RTPs). RTPs are long-term blueprints of regional transportation systems. They analyze transportation needs of metropolitan regions and prioritize project funding decisions (see Figure 6.1).

For this part of our analysis, we reviewed 38 selected RTPs, from the universe of more than 100 plans (see Table 6.1) for qualitative information on goals, objectives, policies, strategies, priorities, and other elements that could be used to analyze conventional planning practices and identify best practices. In nearly all cases, we looked at RTPs for each region in different years to spot trends. The older plans most often provide examples of "conventional practices." The newer plans most often provide examples of "best practices." Results of our qualitative review are summarized in this chapter.

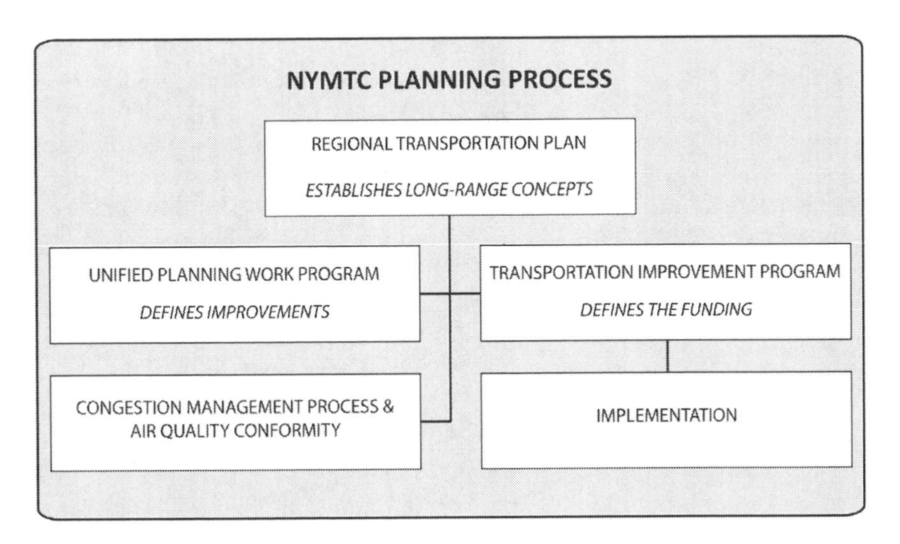

Figure 6.1 MPO Planning Responsibilities.
(New York Metropolitan Transportation Council, 2013, p. 1–28).

Table 6.1 Featured Urbanized Areas and MPOs

Akron, OH	*Akron Metropolitan Area Transportation Study*
Albany, NY	Capital District Transportation Committee
Albuquerque, NM	Mid-Region Metropolitan Planning Organization
Allentown, PA	Lehigh Valley Planning Commission
Atlanta, GA	Atlanta Regional Commission
Austin, TX	Capital Area Metropolitan Planning Organization
Bakersfield-Kern, CA	Kern Council of Governments
Baltimore, MD	Baltimore Regional Transportation Board
Baton Rouge, LA	Capital Region Planning Commission
Berks County, PA	Berks County Planning Commission
Berkshire County, MA	Berkshire Regional Planning Commission
Birmingham, AL	Birmingham Metropolitan Planning Organization
Bismarck, ND	Bismarck-Mandan Metropolitan Planning Organization
Boise, ID	Community Planning Association of Southwest Idaho
Boston, MA	Boston Region Metropolitan Planning Organization
Bridgeport-Stamford, CT	Greater Bridgeport Regional Council
Buffalo, NY	Greater Buffalo Niagara Regional Transportation Council
Burlington, VT	Chittenden County Metropolitan Planning Organization
Cape Coral, FL	Lee County Metropolitan Planning Organization
Charleston, SC	Berkeley-Charleston-Dorchester Council of aaGovernments
Charlotte, NC	Charlotte Regional Transportation Planning Organization
Chattanooga, TN	Chattanooga-Hamilton County Regional Planning Agency
Chicago, IL	Chicago Metropolitan Agency for Planning
Cincinnati, OH	Ohio-Kentucky-Indiana Regional Council of Governments
Cleveland, TN	Cleveland Urban Area Metropolitan Planning Organization
Colorado Springs, CO	Pikes Peak Area Council of Governments
Columbia, SC	Central Midlands Council of Governments
Columbus, OH	Mid-Ohio Regional Planning Commission
Dallas-Ft. Worth, TX	North Central Texas Council of Governments
Dayton, OH	Miami Valley Regional Planning Commission
Denver, CO	Denver Regional Council of Governments

Akron, OH	*Akron Metropolitan Area Transportation Study*
Des Moines, IA	Des Moines Area Metropolitan Planning Organization
Dover, DE	Dover/Kent County Metropolitan Planning Organization
Durham, NC	Durham-Chapel Hill-Carrboro Metropolitan Planning Organization
El Paso, TX	El Paso Metropolitan Planning Organization
Fargo, ND	Fargo-Moorhead Metropolitan Council of Governments
Flagstaff, AZ	Flagstaff Metropolitan Planning Organization
Fresno, CA	Fresno Council of Governments
Ft. Lauderdale, FL	Broward Metropolitan Planning Organization
Grand Rapids, MI	Grand Valley Metropolitan Council
Greenville, SC	Greenville County Planning Commission
Harrisburg, PA	Lebanon County Metropolitan Planning Organization
Hartford, CT	Capitol Region Council of Governments
Honolulu, HI	Oahu Metropolitan Planning Organization
Houston, TX	Houston Galveston Area Council
Indianapolis, IN	Indianapolis Metropolitan Planning Organization
Jacksonville, FL	North Florida Transportation Planning Organization
Kansas City, MO	Mid-America Regional Council
Knoxville, TN	Knoxville Regional Transportation Planning Organization
Lancaster, PA	Lancaster County Planning Commission
Lansing, MI	Tri-County Regional Planning Commission
Las Vegas, NV	Regional Transportation Commission of Southern Nevada
Little Rock, AR	Metroplan
Los Angeles, CA	Southern California Association of Governments
Louisville-Jefferson County, KY	Kentuckiana Regional Planning and Development Agency
Madison, WI	Madison Area Transportation Planning Board
McAllen, TX	Hidalgo County Metropolitan Planning Organization
Miami, FL	Miami-Dade Metropolitan Planning Organization
Milwaukee, WI	Southeastern Wisconsin Regional Planning Commission
Minneapolis, MN	Metropolitan Council
Nashville, TN	Nashville Area Metropolitan Planning Organization

(*Continued*)

Table 6.1 (Continued)

Akron, OH	*Akron Metropolitan Area Transportation Study*
Newark, NJ	North Jersey Transportation Planning Authority
New Haven, CT	South Central Region Council of Governments
New York, NY	New York Metropolitan Transportation Council
Oklahoma City, OK	Association of Central Oklahoma Governments
Omaha, NE	Metropolitan Area Planning Agency
Orlando, FL	MetroPlan Orlando
Palm Beach, FL	Palm Beach Metropolitan Planning Organization
Pensacola, FL	Florida-Alabama Transportation Planning Organization
Philadelphia, PA	Delaware Valley Regional Planning Commission
Phoenix, AZ	Maricopa Association of Governments
Pittsburgh, PA	Southwestern Pennsylvania Commission
Port St. Lucie, FL	Martin Metropolitan Planning Organization
Portland, ME	Portland Area Comprehensive Transportation System
Portland, OR	Metro
Providence, RI	State Planning Council
Provo-Orem, UT	Mountainland Metropolitan Planning Organization
Raleigh, NC	Capital Area Metropolitan Planning Organization
Reno, NV	Regional Transportation Commission of Washoe County
Richmond, VA	Richmond Area Metropolitan Planning Organization
Rochester, NY	Genesee Transportation Council
Sacramento, CA	Sacramento Area Council of Governments
St. Louis, MO	East-West Gateway Council of Governments
Salt Lake City-Ogden, UT	Wasatch Front Regional Council
San Antonio, TX	Alamo Area Metropolitan Planning Organization
San Diego, CA	San Diego Association of Governments
San Francisco-Oakland, CA	Metropolitan Transportation Commission
Santa Barbara, CA	Santa Barbara County Association of Governments
Sarasota-Bradenton, FL	Sarasota/Manatee Metropolitan Planning Organization
Savannah, GA	Coastal Region Metropolitan Planning Organization
Scranton, PA	Lackawanna County Planning Commission
Seattle, WA	Puget Sound Regional Council
Spokane, WA	Spokane Regional Transportation Council
Springfield, MA	Pioneer Valley Metropolitan Planning Organization

Akron, OH	*Akron Metropolitan Area Transportation Study*
Stockton, CA	San Joaquin Council of Governments
Tallahassee, FL	Capital Region Transportation Planning Agency
Tampa-St. Petersburg, FL	Hillsborough Metropolitan Planning Organization
Toledo, OH	Toledo Metropolitan Area Council of Governments
Tucson, AZ	Pima Association of Governments
Tulsa, OK	Indian Nations Council of Governments
Virginia Beach, VA	Hampton Roads Transportation Planning Organization
Washington, DC	National Capital Region Transportation Planning Board
Wichita, KS	Wichita Falls Metropolitan Planning Organization
Winston-Salem, NC	Winston-Salem Urban Area Metropolitan Planning Organization
Worcester, MA	Central Massachusetts Metropolitan Planning Organization
Youngstown, OH	Eastgate Regional Council of Governments

Sample selection

RTPs are usually hundreds of pages in length, often in multiple volumes with multiple appendices. For this reason, we sampled RTPs from the universe of plans rather than reviewing them all.

Whereas we sampled a wide range of urbanized areas, our sampling strategy favored large MPOs over small ones because an initial review suggested that they face bigger challenges and arrive at more innovative solutions. Our sampling strategy sought representation from all parts of the country, both for political and methodological reasons. We wanted to be sure our results could not be dismissed as atypical or biased toward one set of conditions or solutions over others. For the same reason, our sampling strategy sought representation from older, more compact regions as well as newer, more sprawling regions.

Mission/vision

A mission statement is a written declaration of an organization's core purpose and focus that normally remains unchanged over time. A vision statement is aspirational; it lays out the most important primary goals of an organization (Arline, 2014). As an organization grows, its goals may change. Therefore, the vision statement may be revised as needed to reflect the changing environment as goals are met.

When included, the mission and vision of an RTP explicitly state the guiding philosophy of the plan. Specific objectives and selected projects may be traced back to the mission and vision.

Conventional practice

In conventional practice, RTPs contain no underlying mission or vision statements that guide the planning process. Instead, they are collections of generalized goals and grab bags of projects. According to the U.S. Department of Transportation, RTPs play a critical role in the planning process, but they are often

> either providing a direction that is not embraced and implemented by partner agencies, elected officials and the public; instead, piecing together modal and jurisdictional plans without providing a coherent regional and system level direction, or missing the opportunity to consider transportation within a broader context that can include economic development, land use, energy, or environmental concerns.
>
> (USDOT/Volpe Center, 2012, p. 1)

Best practice

In best practice, plans are unified by philosophies that make sense in light of resource constraints and regional goals. One such philosophy-driven plan is the San Francisco Bay Area's, which emphasizes compact development, pricing, and technology:

> We learned that infrastructure investments produce only modest tangible effects at the regional level, and that aggressive pricing and land-use strategies exert much greater influence than transportation projects alone in moving us toward achievement of the performance objectives. We also learned that we must rely on technological innovations to make significant headway toward getting us within range of our goals.
>
> (Metropolitan Transportation Commission, 2009, p. 12)

Another example is San Diego's emphasis on smart growth, in both the 2007 plan and the 2011 plan:

> if we have learned anything in the last decade, it's that we can't build our way out of traffic congestion. This leaves us at a crossroads – the road less traveled may hold the key to how we commute in the future . . . In this era of budget and infrastructure deficits, the ultimate success of this Plan will be measured by how well we implement smart growth as our communities are developed and redeveloped over time. To this end, the 2030 RTP helps strengthen the land use – transportation connection and offers regional transportation funding incentives to support smarter, more sustainable land use.
>
> (San Diego Association of Governments, 2007, p. 1–1)

A final example is St. Louis' 2007 emphasis on "fix it first" and operational improvements over capacity expansion:

> The focus is no longer on building large projects to increase the capacity of roads and highways. Preserving and maintaining the transportation system is where most of the money is spent. From the 1950s through the mid-90s, the purpose of federal, state and to some degree local transportation investments was to build the infrastructure that was adequate and appropriate to keep the nation's economy moving and to preserve the national defense. At the end of the 20th Century, the Interstate was complete. Most of the state, county and local roads needed to support mobility had been constructed. Some major projects remained on the region's wish list, but the overall emphasis shifted from building new roadways to improving the condition, operation, safety and accessibility of the existing system. That will continue for the next 25 years. It may take a seismic shift in thinking to adapt an existing transportation system to the 21st Century when many of its component parts were built 10, 25 and 50 years ago.
>
> (East-West Gateway Council of Governments, 2007, p. 7)

Goals

Background

Goals in an RTP commonly address broad topics ranging from traffic safety to congestion management. The traditional focus on vehicle speeds and mobility is shifting toward accessibility and sustainability through goals that address active travel, greenhouse gas reduction, and energy security.

As an alternative to the speed-centric goal of mobility, accessibility focuses on the ease with which desired activities can be reached from any location (see Chapter 2). Accessibility is the goal of most trips, rather than travel or mobility for its own sake. The more activities available from a given location within a given travel time, the better is the accessibility of that location. Thus, accessibility is a function of both land use patterns and the transportation system that serves them. Clustering common destinations, for example, can improve accessibility by reducing physical travel distances (Litman, 2011). Likewise, mobility substitutes such as telecommunications and delivery services improve accessibility by bringing opportunities directly to a given location (Litman, 2015). Efficient travel decisions such as optimal scheduling and trip chaining also offer accessibility benefits, reducing travel time and/or distance between desired activities (Recker, Chen, & McNally, 2001). Improving accessibility can also relate to other goals such as public health and sustainability through impacts on levels of walking and bicycling. In short, moving beyond a singular focus on speed will require goals that address accessibility over mobility.

Although related to accessibility, sustainability goals encompass an even broader range of concerns. With origins in the environmental movement, sustainable development "meets the needs of the present without compromising the ability of future generations to meet their own needs" (Brundtland, 1987, p. 41. It does so by conserving natural resources and protecting the natural environment. In the transportation sector, the principal threats to sustainability are excessive fossil fuel consumption and the air pollution that results from it. Both are positively correlated with vehicle miles traveled (VMT), vehicle trip rates, and congestion levels because "cold starts," "hot soaks," and low operating speeds contribute to air pollution levels and fuel consumption.

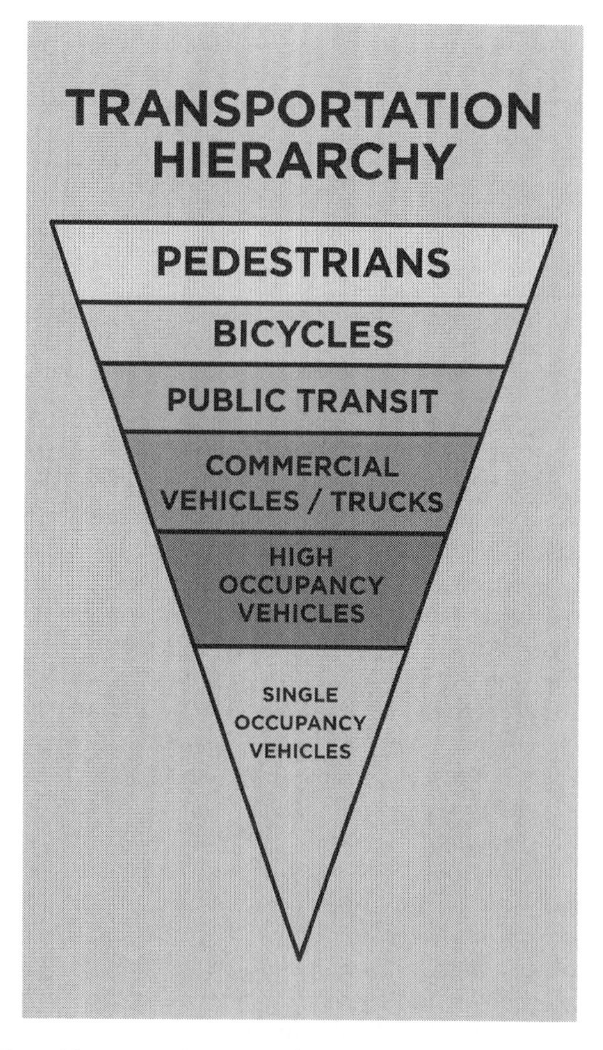

Figure 6.2 Green Transportation Hierarchy.
(City of Portland (OR) Archives, 720002, Climate Action Plan, 2009, p. 36).

A more sustainable transportation system prioritizes use of scarce resources, such as road space and fossil fuels, to favor higher-value trips and more efficient modes over lower-value and less efficient travel, creating a sustainable transportation (or "green") hierarchy. Among travel modes, walking and biking rank highest on the sustainability scale, being affordable, resource-efficient, healthy, and nonpolluting. The single-occupant automobile ranks lowest. Transit and ridesharing may rank high or low, depending on how these modes are accessed and how many seats are occupied.

Conventional practice

In conventional practice, many RTPs focus on the traditional goal of the traffic engineering profession, traffic congestion relief. Whereas they might also acknowledge MAP-21's eight metropolitan planning factors, and often include some discussion of traffic safety, air quality, and environmental justice, there is no doubt where they place their emphasis. The RTPs contain entire chapters on congestion management and lists of highway capacity projects aimed at moderating congestion.

To some degree MPOs are following precedent or habit, and to some extent they are conforming to federal regulations. MAP-21 and its predecessor acts require larger MPOs to adopt a Congestion Management Process with which they monitor mobility within the region, obtain timely information about transportation system performance, and make recommendations to correct deficiencies that are found.

Since the passage of SAFETEA-LU, all states are required to prepare Strategic Highway Safety Plans. RTPs include safety elements that incorporate or summarize the priorities, goals, countermeasures, and projects contained in their respective state Strategic Highway Safety Plans. MPOs' emphasis on air quality is a natural result of the conformity requirements of the Clean Air Act amendments. The recent sensitivity to environmental justice issues is likewise a response to federal law and regulation.

RTPs often devote entire chapters or entire appendices to these regulatory subjects. The Kansas City RTP, for example, has a chapter on air quality conformity (Chapter 11), and appendices on environmental justice (Appendix J) and metropolitan planning factors (Appendix A) (Mid-America Regional Council, 2015). This implies that federal laws and regulations can shift the emphasis toward emerging goals of state or federal policy. The emphasis on congestion management, for example, may tilt the planning process toward capacity expansion, whereas emerging goals may have just the opposite effect, striving instead to minimize highway-induced traffic and contain VMT (vehicle miles traveled).

Best practice

The best practice is to adopt broad goals. Three emerging goals appear in progressive plans. One is to promote active travel. These plans give more emphasis than earlier plans to bicycle and pedestrian facility improvements. Some mention the rise in obesity and the decline in physical activity among Americans.

Portland is a leader in this respect (see Figure 6.3):

> Interest in the connection between urban planning and active living grew in the 1990s, an outcome of a growing interest in "smart growth," a movement to integrate land use, transportation and public health planning. Studies since then report positive effects on human health in neighborhoods built to encourage walking and biking. In addition, transportation systems impact chronic diseases such as asthma that are related to air quality and vehicle emissions. While the Portland region has long embraced such policies, based on land use and transportation benefits, the introduction of health goals and objectives in transportation planning and the RTP is a new realm for the region. There is ample evidence that transportation and community design are critical factors in determining whether residents are able to be physically active enough to ensure their health. The region's transportation system is incomplete from the perspective of enabling sufficient physical activity.
>
> (Portland Metro, 2010, p. 1–32)

The 2011 San Diego plan includes $2.5 billion through 2050 for active transportation. That is more than 2 percent of the total costs of the plan. The rationale:

> Making bicycling and walking viable options for everyday travel can increase mobility, reduce greenhouse gases, and improve public health.
>
> (San Diego Association of Governments, 2011, p. 1–10)

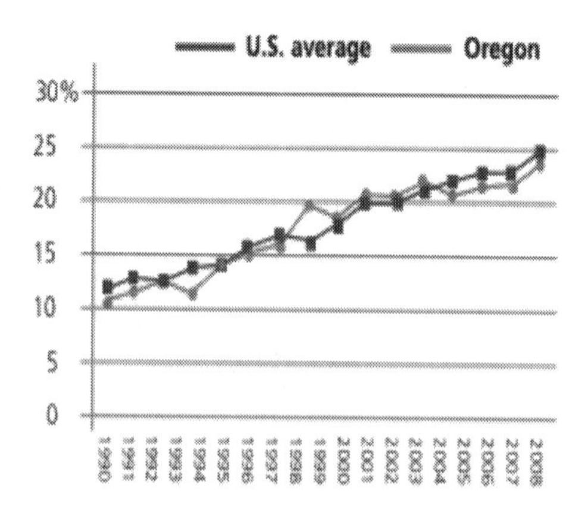

Figure 6.3 Percentage of Adults Who Are Obese, Oregon vs. United States.
(Portland Metro Regional Government, 2010, p. 1–32)

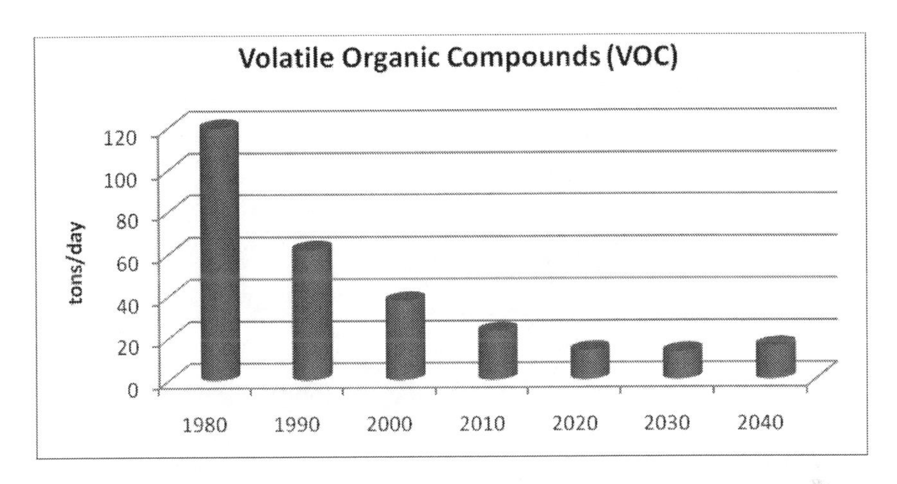

Figure 6.4 Regulated Pollutant Emissions on the Decline (Unlike CO_2 Emissions). (Wasatch Front Regional Council, 2011, p. 258).

Another emerging goal is to mitigate greenhouse gas emissions. This is different than the historical emphasis on regulated pollutants. Climate goals are less susceptible to improvements in vehicle technology (see Figure 6.4), as CO_2 emissions cannot be captured at the tailpipe. The level of greenhouse gas emissions is far more dependent on the amount of driving and aggregate VMT.

As the fingerprints of climate change become undeniable, and pressure from other countries mounts, Congress will eventually pass climate legislation. When this occurs:

> it is highly likely the Commission will need to identify new GHG reduction goals, targets, milestones and reduction actions. While these new requirements are not yet known, the scientific evidence is overwhelming and need for immediate global response is compelling.
>
> (Tri-County Regional Planning Commission, 2010, p. 1–8)

In part because its plan was adopted at a time of heightened concern about climate change (see Figure 6.5), and in part because it is located in the leading state for climate action, the San Francisco RTP gives at least as much emphasis to climate mitigation as to the conventional goal of congestion relief. Not only does its plan contain a whole section on climate ("Lead the Charge on Climate Protection"), but it establishes a new multimillion dollar Climate Grants Program to determine what strategies can most effectively reduce GHG emissions (Metropolitan Transportation Commission, 2009, p. 46). Potential projects may seek to increase the use of low-GHG alternative fuels, expand car-sharing programs, or implement pricing demonstration projects.

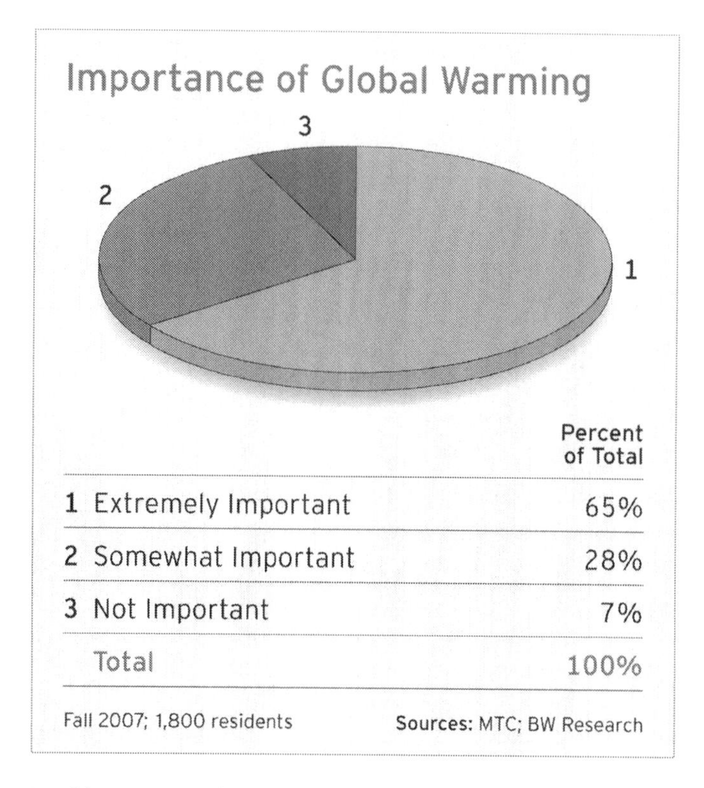

Figure 6.5 Public Concern about Climate Change in the Bay Area.
(Metropolitan Transportation Commission, 2009, p. 19).

> Our transportation decisions and actions can either help or hinder efforts to protect the climate, and to this end, the Commission has set aside $400 million to implement a Transportation Climate Action Campaign that focuses on individual actions, public-private partnerships, and incentives and grants for innovative climate strategies. Known for its commitment to the environment, the Bay Area is ideally suited to provide regional leadership and serve as a model for California, the nation and the world in our efforts to reduce our carbon footprint.
>
> (Metropolitan Transportation Commission, 2009, p. 14)

The third emerging goal is energy security in the face of so-called peak oil. Researchers now are engaged in estimating when the phenomenon will occur at a global scale. Peak oil production ("peak oil") occurred at the national level in countries around the world, with peaks in the lower 48 United States, Alaska, and Mexico occurring in 1971, 1989, 2004, respectively (Zittel & Schindler, 2007). The rise of hydraulic fracturing (fracking) has extended the date of peak oil, but it has to occur at some point (see Figure 6.6). Moreover, the environmental consequences of fracking are just beginning to be understood, and these could put a brake on fracking activities. Oklahoma, for example, went from experiencing fewer than two quakes of magnitude 3 or greater per year to 907 quakes of magnitude 3 and above in 2015 (Wines, 2016). These earthquakes have been largely human-caused. When naturally occurring wastewater released during the fracking process is pumped back into the ground, it increases pressure on subterranean faults, causing them to slip and produce tremors. Oklahoma is now the most earthquake-prone state in the country behind only Alaska (and ahead of California).

Another environmental concern associated with fracking is groundwater pollution. In Wyoming, the town of Pavillion reported benzene levels 50 times higher than the allowable limit under EPA standards. The presence of other contaminants in the water supply was elevated relative to EPA standards as well. DiGuilio and Jackson (2016) established a link between this localized water pollution and fracking practices in the area.

Concerns about fracking could impact the long-term viability of the practice in the United States. Oklahoma has already taken steps to limit the

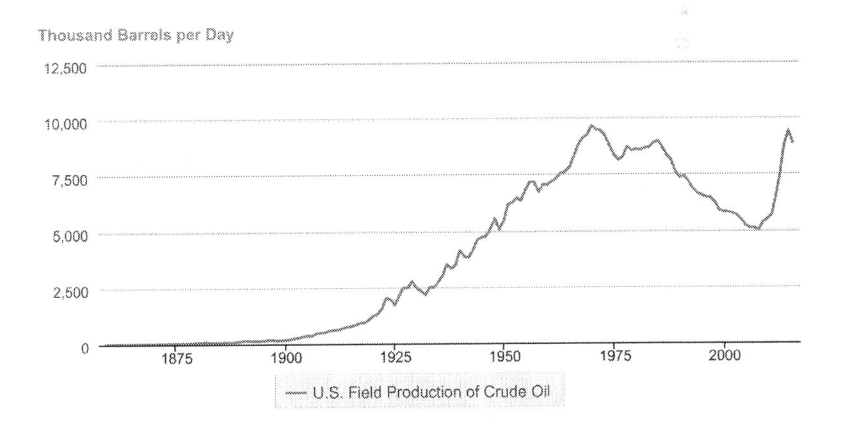

Figure 6.6 U.S. Production of Crude Oil.

(U.S. Energy Information Administration, 2017; Retrieved from www.eia.gov/dnav/pet/hist/LeafHandler.ashx?n=PET&s=MCRFPUS2&f=A).

amount of wastewater injected into some wells, although the EPA has called on the state to do more (AP Staff, 2017). Because of these concerns, fracking should not be seen as a solution to energy independence. Rather, we need to find ways to reduce transportation energy consumption in the longer term, potentially even within the 30-year timeframe of the typical RTP.

Policies designed to mitigate climate impacts by reducing VMT will also increase energy security.

> Over the coming decades, the profound transformation of the global economy to use less energy and produce less GHG presents a tremendous opportunity for Greater Philadelphia . . . . As energy prices increase and governmental policies to curb GHG emissions are put in place across the nation and world, businesses and individuals will increasingly select places where they can meet their needs with less energy from fossil fuel sources. In order to keep the DVRPC region competitive, a multipronged strategy is needed to effectively address both GHG emissions and energy use.
>
> (Delaware Valley Regional Planning Council, 2009, p. 78)

Objectives and performance measures

Background

When the goals of an RTP shift from mobility and congestion management to address issues of accessibility and sustainability, a different set of performance measures is required. For example, there is considerable evidence that designing roads to the highest automobile service standards may not be the best way to lower vehicle emissions and increase a network's sustainability. When accounting for feasibility and marginal improvements, for example, level of service C has been shown to be the most effective operating point in terms of emissions reductions (Cobian et al., 2009.

Emerging areawide measures now target roadway connectivity or pedestrian route connectivity (Steiner, Bond, Miller, & Shad, 2004). National leaders in growth management utilizing areawide level-of-service measures include metropolitan areas in Florida; Montgomery County, Maryland; San Jose, California; and Bellevue, Washington. Since 1999, Florida has applied areawide level-of-service measures to multiple modes, with criteria for pedestrians, bicycles, and transit, in addition to motor vehicles.

Vehicle miles traveled (VMT) reduction targets have also been used in some jurisdictions as part of overall efforts to implement transportation demand management, and to achieve specific planning objectives such as congestion and emission reductions (Litman, 2013). This class of performance measures includes vehicle trip rates, vehicle miles traveled (VMT), and VMT on a per-person or per-household basis. VMT also provides a primary performance measure in the U.S. Green Building Council's sustainable neighborhood development rating system (Ewing et al., 2013).

One thing VMT does not measure is congestion, which along with VMT, is a major determinant of vehicle emissions, fuel consumption, and time wasted in traffic. It has been suggested that vehicle hours traveled (VHT) might be a better measure of travel demand than VMT, at least for air quality planning purposes because vehicle emissions per mile decline with increasing speed (up to about 45 mph) whereas vehicle emissions per hour are essentially independent of speed. Transportation control strategies that reduce VMT may or may not improve air quality, depending on their effect on travel speeds. Alternatively, strategies that reduce VHT can be consistently depended on to improve air quality.

Conventional practice

All RTPs provide information on inputs, specifically dollars to be spent, but many fail to fully quantify what is accomplished with all those billions of dollars. In conventional practice, if performance measures were provided, the focus was on congestion, and other performance outcomes were defined only in qualitative terms. The 2008 Atlanta RTP contained hundreds of qualitative objectives, but there was only one quantitative objective, and it related to congestion.

> The region's TTI [travel time index – free flow travel time divided by free flow travel time] system-wide target is 1.35. A value of 1.35 is equivalent to traffic reaching approximately 67% of designed capacity. Generally, 67% of designed capacity equates to receiving a mobility grade of "C." Rush hour traffic typically rises to level "E" or "F." A grade of "F" is the worst possible grade and reflects gridlock conditions. A grade of "A" represents the most desirable traffic conditions (free flow). In order to capture a better understanding of where the worst congestion is, the CMP includes a second benchmark beyond the 1.35 target with a TTI value of 1.80. This value represents the type of congestion typically experienced during weekday afternoons and is equivalent to an "E" or "F."
>
> (Atlanta Regional Commission, 2008, p. 24)

In the Atlanta RTP, nearly all of the performance comparisons, both in the text and in graphs, related to congestion. Notwithstanding this emphasis on congestion relief, traffic conditions in Atlanta were projected to get dramatically worse.

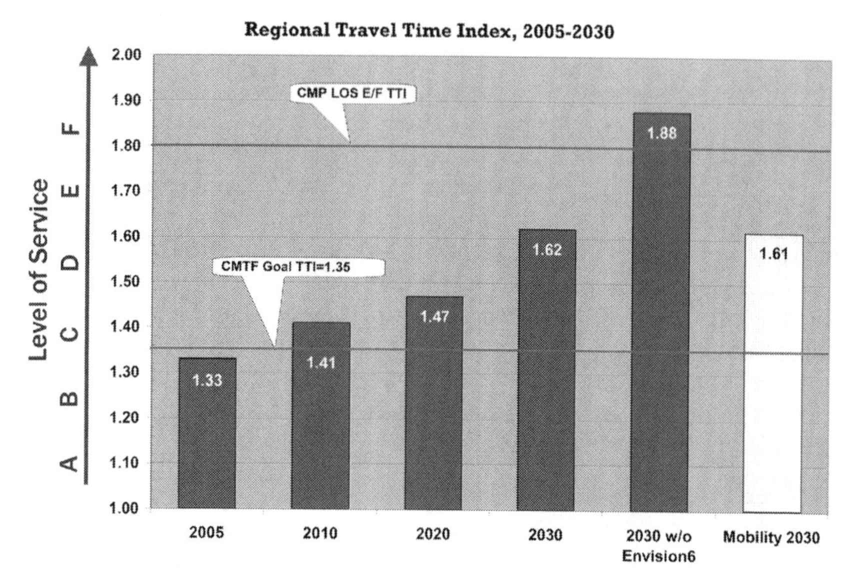

Figure 6.7 Regional Travel Time Index, 2005–2030.
(Atlanta Regional Commission, 2008, p. 78).

Best practice

"What gets measured gets done." "If you don't know where you're going, all roads lead there." These and other old saws argue for a balanced scorecard of measurable performance objectives. The Portland RTP contains such a scorecard:

- Safety – By 2035, reduce crashes, injuries and fatalities by 50 percent compared to 2005.
- Congestion – By 2035, reduce vehicle hours of delay per person by 10 percent compared to 2005.
- Climate change – By 2035, reduce carbon dioxide emissions by 40 percent below 1990 levels.
- Active transportation – By 2035, triple walking, biking and transit trips compared to 2005.
- Clean air – By 2035, ensure zero percent population exposure to at-risk levels of air pollution.
- Travel – By 2035, reduce vehicle miles traveled per person by 10 percent compared to 2005.
- Affordability – By 2035, reduce the average household combined cost of housing and transportation by 25 percent compared to 2000.
- Access to daily needs – By 2035, increase by 50 percent the number of essential destinations accessible within 30 minutes by bicycling and

> public transit for low-income, minority, senior and disabled populations compared to 2005.
>
> (Portland Metro Regional Government, 2010, p. 2–13)

Significantly, performance measures in the same plan suggest that Portland will fall short of several of its objectives for 2035. That is useful information to have as Portland updates its RTP in the years to come.

Demographics

Background

There is an expression, "demographics are destiny." Between now and 2050, the United States will experience dramatic growth in its older population. In 2050, the population aged 65 and over is projected to be 83.7 million, almost double its estimated population of 43.1 million in 2012. The baby boomers are largely responsible for this increase.

As one RTP notes in a chapter on the "Land Use-Transportation Connection: Growing Smarter," this demographic trend has implications for regional land use and transportation planning.

> Key among these trends is the aging of the population. This trend has widespread implications and points to the need for communities with a wider variety of housing choices, more affordability, more accessible public transportation, more walkability, and a greater mix of land uses ensuring the appropriate public facilities needed to serve the changing demographics.
>
> (San Diego Association of Governments, 2007, p. 5-1)

Conventional practice

Nearly all RTPs contain population and employment forecasts, and many include forecasts for age cohorts. Those that do all reach the same conclusion: that the most significant demographic shift from the base to the horizon year will be the aging of the population, as in Atlanta (see Figure 6.8).

> The 20-county region is forecast to grow older over the next three decades . . . Between 2010 and 2040, growth among the senior population (65 years and older) will comprise 39% of overall population growth within the 20-county region. The proportion of senior residents in the 20-county region is forecasted to grow from 8.9% in 2010 to 18.8% of the total population in the 20-county region by 2040.
>
> (Atlanta Regional Commission, 2014, p. 2–14)

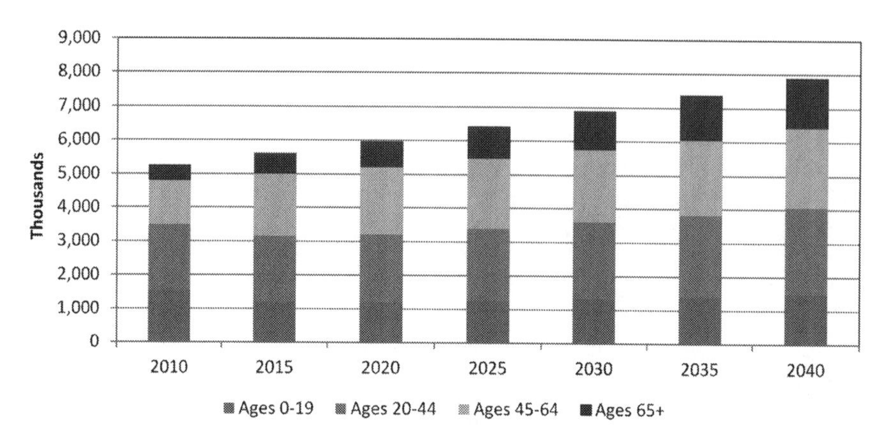

Figure 6.8 Changing Age Distribution.
(Atlanta Regional Commission, 2014, p. 2–14).

While acknowledging the trend, few RTPs fully explore the implications for housing choices and travel needs. Typical is this status quo modal assessment from the Cincinnati RTP:

> While the proportion of the population in their retirement years will be increasing during the planning period, the proportion of this age group that drives will increase even faster. As today's population ages, the elderly of the future will be almost universally licensed to drive, accustomed to driving on a nearly daily basis and concentrated in suburban areas that are auto-dependent. The elderly will continue to drive as long as they are physically or legally able.
>
> (Ohio-Kentucky-Indiana Council of Governments, 2012, p. 3–4)

Typical of the housing market analyses is this status quo assessment from the Philadelphia RTP:

> Many aging suburban baby boomers will want to stay in the suburban communities in which they have raised their families after they retire. Challenges facing seniors include a lack of affordable and accessible housing alternatives; limited accessibility within their existing homes; limited accessibility within their communities; and difficulties with transportation and mobility, especially given the lack of public transit in many suburban locations. Elderly homeowners will also face economic challenges, as the cost of essentials such as transportation and health care skyrocket, leaving less money available for housing costs (including

> rising property taxes and the costs of home repair and maintenance). It is imperative that the region's elected officials, planners, service providers, and the elderly and near-elderly themselves plan now to accommodate the coming "senior boom."
>
> (Delaware Valley Regional Planning Council, 2009, p. 19)

The response to demographic change is usually reactive rather than proactive, in keeping with the notion that MPOs have no control over land use patterns.

> Starting in 2010, baby boomers will start turning 65. By 2035, over half a million people in the region will reach retirement age and many of them will leave the workforce. The expected jobs/labor force imbalance will increase the distance people will travel for work, as well as produce longer commute times and more congested roadways.
>
> (Baltimore Regional Transportation Board, 2007, p. 86)

Best practice

The best practice would not only acknowledge this revolutionary change in the age distribution, but would fully explore the implications for land use and transportation planning. Among the implications are a slowdown in the growth of VMT and a growing demand for community-level transit transportation network services Uber and Lyft, and housing in dense, walkable activity centers. Certain RTPs deal with these realities in qualitative terms, but few quantify and plan for the effects.

From the San Francisco RTP comes this qualitative assessment of changing travel and housing demands:

> Key among the demographic changes that will affect Bay Area transportation is the aging of the Baby Boomers. As this sizeable segment of the region's residents reaches senior status, it is expected that many will relocate into smaller dwellings in the more urban portions of the Bay Area to have easier access to essential services and cultural opportunities. For some, with aging will come a loss of the ability to drive, and for those with low incomes or physical disabilities, 'lifeline' transportation issues will become increasingly important. From a land use and mobility perspective, then, the graying of the Baby Boomers would seem to argue for a greater emphasis on smaller homes, low maintenance housing arrangements, and a heavier reliance on non-driving transportation options, such as transit and ride-sharing with younger friends and family.
>
> (Metropolitan Transportation Council, 2009, p. 8)

Probably the main implication of this demographic shift is the need for smart growth initiatives like those described below, fostering dense, mixed-use centers where seniors can age comfortably in place and remain mobile after they cease driving.

Scenario planning

Background

Alternative regional development patterns have been compared in simulation studies, through a process of scenario planning (see Chapter 4). Some patterns balance jobs and housing by sub-region, others mix jobs and housing without striving for balance, and still others segregate jobs and housing into large single-use districts. Some concentrate jobs in the urban core of the region, others in a few major centers, and still others in a multitude of smaller centers. The best alternatives in terms of both congestion and VMT seem to be those that concentrate development to a degree (but not too much) and balance jobs and housing within sub-areas (Cervero & Duncan, 2006; Bartholomew & Ewing, 2009). These scenarios offer a number of benefits for many regions, such as the reduced VMT offered by Boise, Idaho's smart growth regional transportation plan alternative or the improved density offered by Sacramento's preferred blueprint scenario.

Conventional practice

The conventional practice is to test future alternative transportation investments against a fixed future land use forecast, to see how well the former meets the travel demands generated by the latter. In starting its scenario planning process, the Boise MPO noted:

> past plans started with a single view of future growth and became a process of asking participants what transportation projects they wanted. The resulting lists were assembled into a plan.
>
> (Community Planning Association of Southwest Idaho, 2006, p. 26)

In conventional practice, future land use patterns are taken as a given that cannot be modified by an MPO or its constituent governments. This is what was done in Cincinnati, where future land use patterns were assumed to be even more sprawling than present day's (see Figure 6.9). A low-density development pattern was assumed despite the fact that real fuel prices are likely to be dramatically higher by 2040 and lifestyle changes are likely to favor dense, mixed-use development (see Figure 6.10).

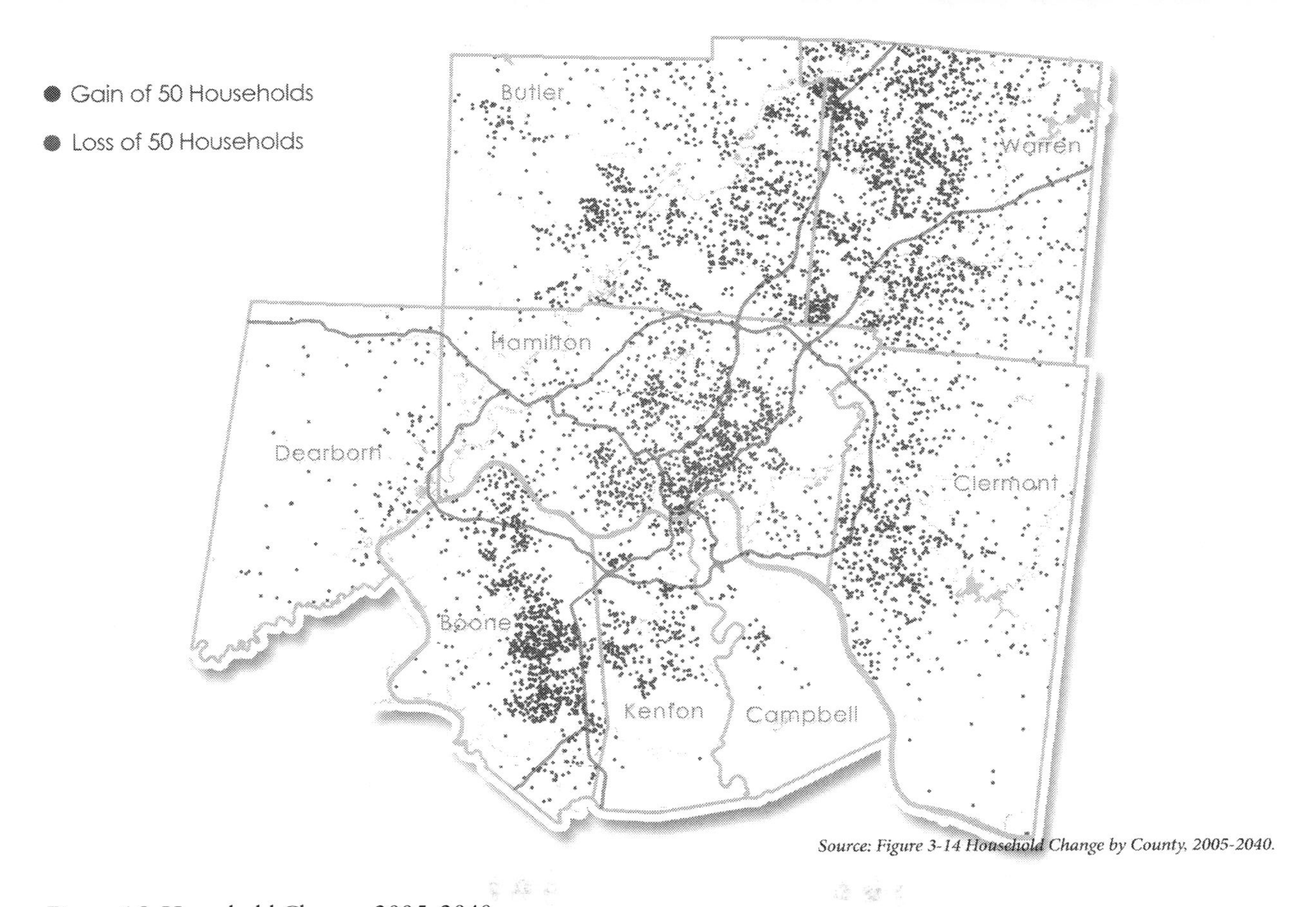

Figure 6.9 Household Change, 2005–2040.
(Ohio-Kentucky-Indiana Council of Governments, 2012, p. 3–9).

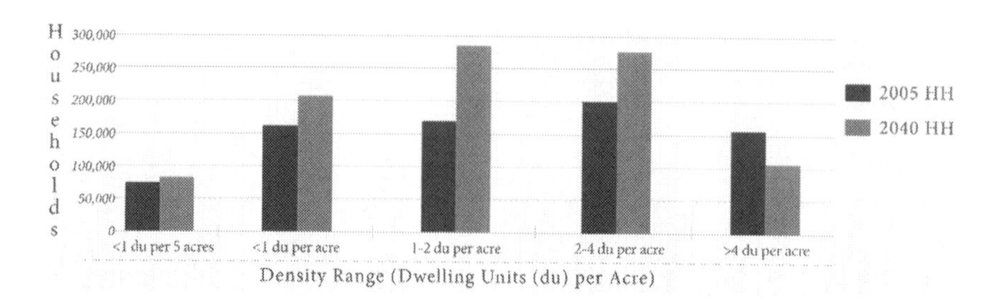

Figure 6.10 Household Distribution by Density, 2005–2040.
(Ohio-Kentucky-Indiana Council of Governments, 2012, p. 3–10).

Best practice

An alternative approach, exemplified by RTPs of Baltimore, Boise, Boston, Colorado Springs, Lansing, Portland, Sacramento, Salt Lake City, and other leading regions, assumes that future development patterns can take various forms, just as future transportation investments can take various forms (Baltimore Regional Transportation Board, 2011; Community Planning Association of Southwest Idaho, 2010; Boston Region Metropolitan Planning Organization, 2011; Pikes Peak Area Council of Governments, 2012; Tri-County Regional Planning Commission, 2010; Portland Metro Regional Government, 2014; Sacramento Area Council of Governments, 2012; Wasatch Front Regional Council, 2015).

These various forms are run through the regional travel demand model to see how well they perform with respect to VMT, congestion, and other transportation outcomes. The main conclusion from these scenario plan studies is that transportation outcomes depend as much on development patterns as they do on transportation system improvements. In the 2005 Burlington RTP:

> the single most important factor to improving transportation system performance is to move toward a land use pattern based on concentrated development similar to that identified in the CCRPC Regional Plan. In fact, the success of the future transportation system investments outlined in the Preferred Alternative above is dependent upon achieving [this] development pattern.
>
> (Chittenden County Metropolitan Planning Organization, 2005, p. 59)

Likewise, in the Lansing RTP:

> TCRPC's primary strategy to address congestion throughout the region is land use based. The primary element of this land use strategy is

> implementing the adopted "Wise Growth" scenario, the regional land use vision developed through the "Regional Growth: Choices for Our Future" project. As noted, implementing this strategy reduces congested lane miles to nearly half of what would otherwise occur at regional build out under densities permitted by current zoning.
>
> (Tri-County Regional Planning Commission, 2010, p. 8–8)

Inherent in scenario planning studies is the belief that future development patterns can be influenced by the MPO, through its investments and the leverage it exercises over local governments through planning and zoning.

The use of scenario planning for positive ends is illustrated in Sacramento. Sacramento's previous RTP had come up short in a number of respects:

> the focus on these indicators began with the adoption of the Metropolitan Transportation Plan for 2025 (MTP2025) in July 2002. Although adopted unanimously by the SACOG Board, the Board was extremely concerned about several worrisome projections presented in that plan:
>
> - **VMT growth continuing to outstrip population growth.** The plan was based on a projected population growth of 49 percent between 2000 and 2025, but VMT was projected to grow by about 65 percent over the same period. This meant that the average household needed to drive 8 percent more vehicle miles in order to live, work and play in the Sacramento region. Given the air quality problems in the region, and the strong relationship between vehicle emissions and VMT, this trend was a great concern.
> - **Roadway congestion growth far in excess of growth in VMT.** Even with all of the investments in the MTP2025, roadway congestion experienced by the average household was expected to increase by 58 percent. Total region-wide VMT on heavily congested roadways was expected to increase by 230 percent.
> - **Transit ridership increases, but not by much.** The region-wide transit mode share was projected to increase from 1.0 to 1.2 percent of all trips, even with a large increase in transit service.
> - **Loss in non-motorized mode share.** The percentage of trips made by bike and walk modes was projected to decrease from 6.9 percent to 6.6 percent.
>
> (Sacramento Area Council of Governments, 2008, p. 45)

Disenchantment with the earlier RTP led to a scenario planning process, and the adoption of a preferred scenario based on seven smart growth principles that outperformed the earlier plan in all respects including a 26 percent reduction in regional VMT (see Figures 6.11a,b).

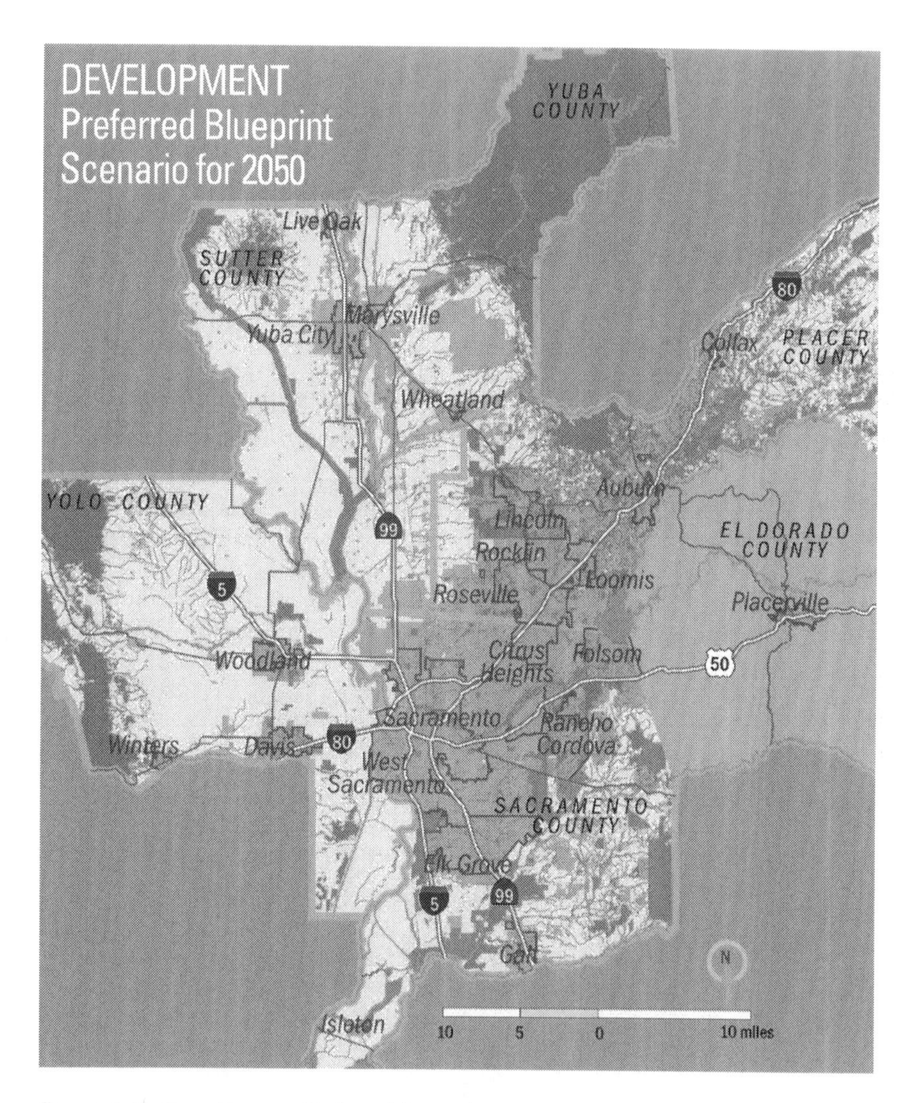

Figure 6.11 Base Case vs. Preferred Scenario.
(Sacramento Area Council of Governments, 2007, pp. 2–3).

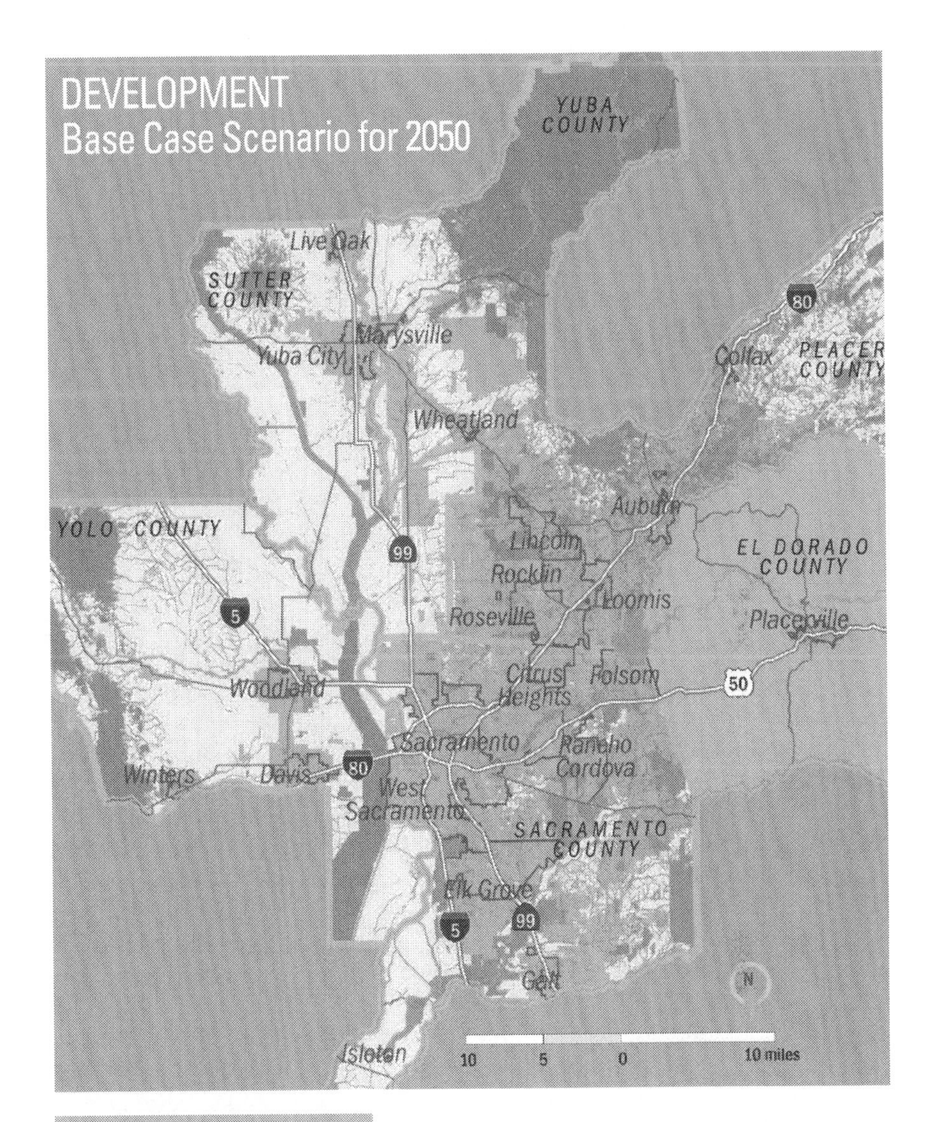

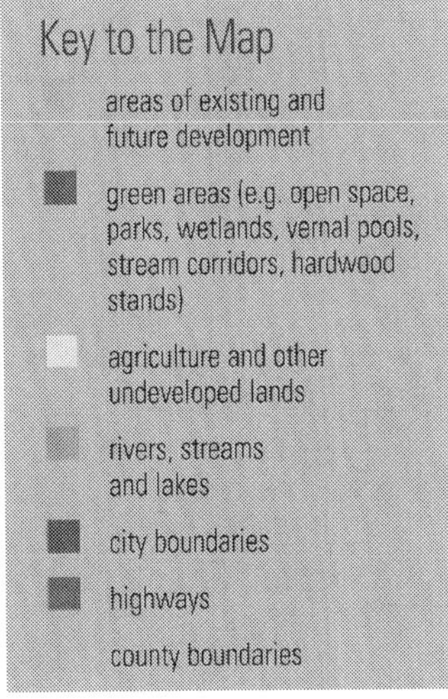

Figure 6.11 (Continued)

Land use-transportation integration

Background

As limitations of mobility measures become more apparent, interest in land planning has grown among transportation professionals. Land planning as frequently practiced today in the United States, where government too often simply reacts to developers' proposals, largely ignores many pressing transportation issues. Increasingly, however, professionals will emphasize proactive land planning, where government seeks to moderate the need for automobile travel.

Ordinarily the largest geographic area of interest in land planning is the region. In regional planning, the focus is on work trips, keeping solo commutes as short as possible and inducing some solo commuters to switch to carpooling or transit. Toward this end, development must be contained rather than allowed to sprawl endlessly; employment must be concentrated in centers; growth must emphasize mixed-use development; and some balance of jobs and housing must be achieved within sub-regions (Cervero & Duncan, 2006; Cervero & Murakami, 2010; Stoker and Ewing, 2014).

RTPs recognize and pursue these planning principles to varying degrees. The majority decline to proactively coordinate land use and transportation, in favor of adapting transportation systems to existing land use plans and conditions.

In contrast to regional planning, local community planning focuses on convenience shopping, recreation, and school trips, such trips being both more frequent and more affected by accessibility considerations than are, say, comparison shopping trips or medical trips. Put another way, residents are more likely to patronize a nearby grocery store than a nearby furniture store, price and quality being more important than accessibility when it comes to big-ticket items like furniture. Also, residents make tens or hundreds of trips to grocery stores for every trip to a furniture store. Thus, from a travel standpoint, we gain much more by "internalizing" grocery stores within communities (Cervero and Duncan, 2006). The goal is to internalize as many of these trips as possible, keeping them off the regional road network. Trips completed entirely within a community tend to be shorter. If trips are short enough, many may switch to walking or bicycling Ewing and Cervero, 2010; Greenwald, 2006; Bhatta & Larsen, 2011;).

Conventional practice

MPOs have historically embraced a hands-off policy when it comes to land use planning, even while acknowledging the critical links between land use and transportation. As the Raleigh and Durham RTPs put it:

> Land use in the Triangle is the responsibility of each local government, not the MPOs. But few things influence the functionality and effectiveness of our transportation system as much as the locations, types,

> intensities and designs of new developments in our region. If we are to successfully provide for the mobility needs of the 1.6 million people here today and the additional 1.3 million expected to be added over the timeframe of this plan, we will need to do a top-notch job of matching our land use decisions with our transportation investments.
>
> (Capital Area Metropolitan Planning Organization and Durham-Chapel Hill-Carrboro Metropolitan Planning Organization, 2013, p. 49)

The Colorado Springs MPO ranked three transportation system scenarios in the following order:

1) Reduced Environmental Impacts System
2) Balanced Investment System
3) Strategic Corridors System

Although the Reduced Environmental Impacts scenario had the highest score, it would have required local governments to change their adopted land use plans to increase density along rapid transit corridors. Advisory committees therefore recommended and the MPO adopted the less preferred alternative because it would not require any changes in local land use plans (Pikes Peak Area Council of Governments, 2008).

Best practice

RTPs routinely call for the integration and coordination of land use and transportation, as in the Honolulu RTP:

> Develop, operate, and maintain Oahu's transportation system in a manner that integrates effective land use and transportation with established sources of funding in a fair and equitable manner.
>
> (Oahu Metropolitan Planning Organization, 2011, p. 11)

Even though the MPO may have no jurisdiction over land use decisions, many RTPs show a trend towards integrating land use and transportation (USDOT/Volpe Center, 2012). But what does integration mean, and how can it be achieved? MPOs can use the power of the purse to affect land use decisions. The Boston MPO is one of several that do so:

> The MPO considers land use and economic development in its project-prioritization and funding processes so that transportation spending will respond both to current conditions and to future needs likely to result from local and regional plans and priorities. The selection process for projects in JOURNEY to 2030 included consideration of land use and

> economic development factors. Those factors are also included in the criteria the MPO uses to select projects for funding in its Transportation Improvement Program (TIP).
>
> (Boston Region Metropolitan Planning Organization, 2007, p. 11–3)

What local decisions can and should MPOs seek to influence? The Raleigh and Durham RTPs provide a partial list:

> The ties between regional transportation interests and local land use decisions are most pronounced in three cases:
>
> 1) Transit Station Area Development.
> 2) Major Roadway Access Management.
> 3) Complete Streets & Context-Sensitive Design.
>
> (Capital Area Metropolitan Planning Organization and Durham-Chapel Hill-Carrboro Metropolitan Planning Organization, 2013, p. 49)

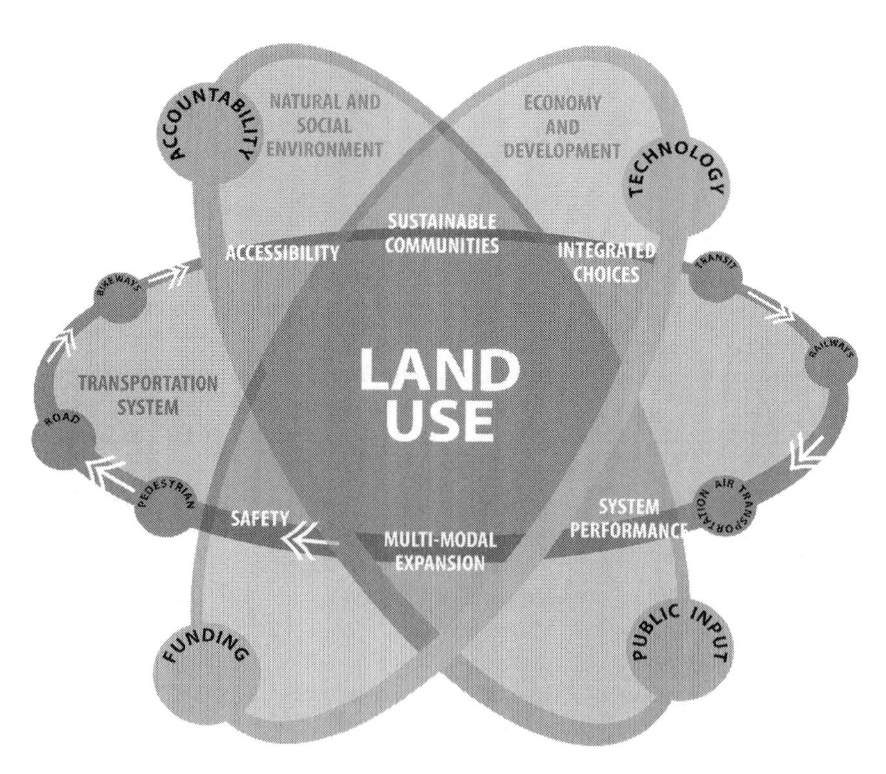

Figure 6.12 Transportation Solutions Revolve around Land Use.
(Pima Association of Governments, 2010, p. 60).

Although most would agree that the ties between land use and transportation transcend these three areas, these are good examples of areas in which MPOs can (though seldom do) take a leadership role.

The Metropolitan Transportation Commission in San Francisco may be the best example of an MPO that has successfully promoted TOD. Resolution 3434 establishes minimum levels of development around transit stations along new transit corridors.

> the Commission's 2005 adoption of the Resolution 3434 Transit-Oriented Development (TOD) Policy helps to maximize the effectiveness and value of regional services by conditioning discretionary funds on transit-supportive land uses. In fact, the TOD policy will help stimulate the construction of at least 42,000 new housing units and boost the region's overall transit ridership by over 50 percent by 2035.
>
> (Metropolitan Transportation Commission, 2009, p. 17)

Probably the best example of an MPO promoting access management is in Reno (see Table 6.2).

> The Access Management Standards shown in (the accompanying table) will be used in the design of future improvements to the Regional Road System (RRS) and the classification of existing improvements for planning purposes.
>
> (Regional Transportation Commission of Washoe County, 2008, p. 2–4)

The best example of an MPO that has weighed in on complete streets and context-sensitive design is in Portland. Portland Metro has published four guidebooks related to street design. *Creating Livable Streets: Street Design Guidelines for 2040* provides model street designs that successfully integrate streets with nearby land uses to enhance safety and promote community livability (Portland Metro Regional Government, 2004). The guidelines have provided the impetus for reform of street standards by the region's localities (see Figure 6.13).

Urban form

Background

The relationship between the built environment and travel is perhaps the most heavily researched subject in urban planning. There are now over 200 empirical studies conducted with a degree of rigor, that is, with decent sample sizes, controls, and statistical tests to determine the significance of the various effects (see reviews by Badoe & Miller, 2000; Crane, 2000; Ewing & Cervero, 2001; Saelens, Sallis, & Frank, 2003; Heath et al., 2006; Ewing & Cervero, 2010; Stevens, 2017; Ewing and Cervero, 2017).

Table 6.2 Access Management Standards

Access Management Standards-Arterials[1] *and Collectors*							
Access Management Class	*Posted Speeds*	*Signals Per Mile and Spacing*[2]	*Median Type*	*Left From Major Street? (Spacing from signal)*	*Left From Minor Street or Driveway?*	*Right Decel Lanes at Driveways?*	*Driveway Spacing*[3]
High Access Control	45-55 mph	2 or less Minimum spacing 2350 feet	Raised w/channelized turn pockets	Yes 750 ft. minimum	Only at signalized locations	Yes[4]	250 ft./500 ft.
Moderate Access Control	40-45 mph	3 or less Minimum spacing 1590 feet	Raised or painted w/turn pockets	Yes 500 ft. minimum	No, on 6-or 8-lane roadways w/o signal	Yes[5]	200 ft./300 ft.
Low Access Control	35-40 mph	5 or less Minimum spacing 900 feet	Raised or painted w/turn pockets or undivided w/ painted turn pockets or two-way, left-turn lane	Yes 350 ft. minimum	Yes	No	150 ft./200 ft.
Ultra-Low Access Control	30-35 mph	8 or less Minimum spacing 560 feet	Raised or painted w/turn pockets or undivided w/ painted turn pockets or two-way left-turn lane	Yes 350 ft. minimum	Yes	No	150 ft./200 ft. 100 ft./100 ft.[6]

(Regional Transportation Commission of Washoe County, 2013, p. E-3)

1 On-street parking shall not be allowed on any new arterials. Elimination of existing on-street parking shall be considered a priority for major and minor arterials operating at or below the policy level of service.

2 Minimum signal spacing is for planning purposes only; additional analysis must be made of proposed new signals in the context of existing conditions, planned signalized intersections, and other relevant factors impacting corridor level of service.

3 Minimum spacing from signalized intersection/spacing from other driveways.

4 If there are more than 30 inbound, right-turn movements during the peak-hour.

5 If there are more than 60 inbound, right-turn movements during the peak-hour.

6 Minimum spacing on collectors.

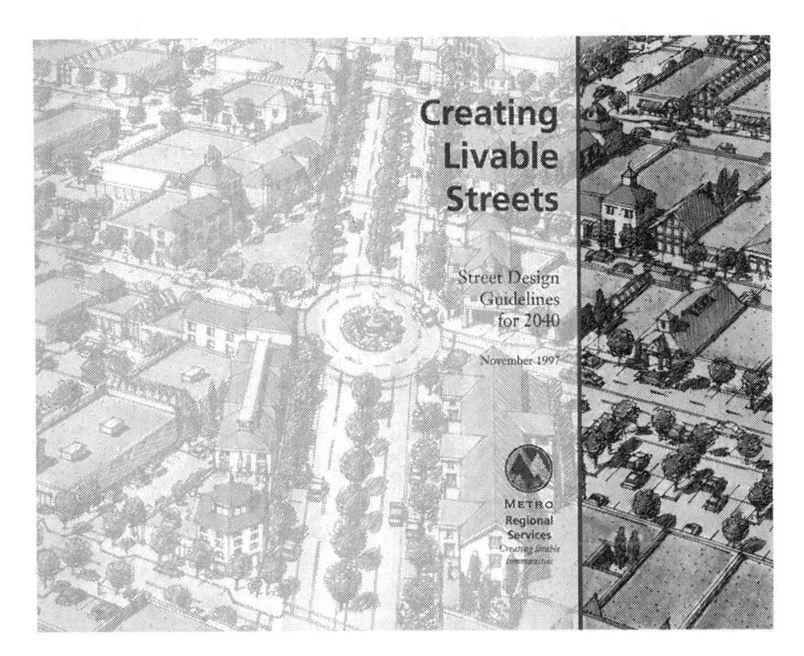

Figure 6.13 Portland Street Design Guidebook.
(Portland Metro Regional Government, 2004).

Nearly all of these studies show significant relationships between travel behavior and the "D" variables: density, diversity, design, destination accessibility, and distance to transit. Density is usually measured in terms of persons, jobs, or housing units per unit area. Diversity refers to land use mix. It is often related to the number of different land uses in an area, and the degree to which they are "balanced" in land area, floor area, or employment. Design includes street network characteristics within a neighborhood. Street networks vary from dense urban grids of highly interconnected, straight streets, to sparse suburban networks of curving streets forming "loops and lollypops."

Destination accessibility refers to a household's ability to reach local or regional attractions such as jobs and shopping. Accessibility tends to be highest in central cities and lowest in peripheral bedroom suburbs. It is measured in terms of distance to closest attractions or weighted distance to attractions regionally. Distance to transit is usually measured in terms of walking distance to bus stops or rail stations. It usually relates to high-quality transit, not regular bus service that has less impact on mode choice and locational decisions.

According to Ewing and Cervero (2010) in their meta-analysis of over 200 built environment and transportation studies, the two measures most strongly associated with VMT are job accessibility by auto and distance to downtown, both of which are proxies for the destination accessibility D variable. The distance to downtown variable is likely correlated with many of the other Ds: living in the core city typically means higher densities in mixed-use settings with good regional accessibility. Next most strongly associated with VMT were design metrics expressed in terms of intersection density or street connectivity. Both short blocks and many interconnections shorten travel distances, apparently to about the same extent.

Among the remaining Ds, density has the weakest association. The relatively weak relationships between density and travel likely reflect density's role as an intermediate variable that ultimately gets expressed by the other Ds – that is, dense settings usually have mixed uses with small blocks and plentiful intersections that shorten trips and encourage walking.

Conventional practice

Just as the emphasis in transportation plans is on roadway congestion, so is the emphasis in associated land use plans on one aspect of urban form, density (see Figures 6.14 and 6.15). From the 2007 Boston RTP:

> Recent travel demand modeling results for the Boston region suggest that changes in land use that create denser future developments located near existing transportation facilities will have a more positive impact on reducing congestion, increasing mobility, and improving air quality than all the new transportation projects the region can afford to build in the next 23 years.
>
> (Boston Region Metropolitan Planning Organization, 2007, p. 11–2)

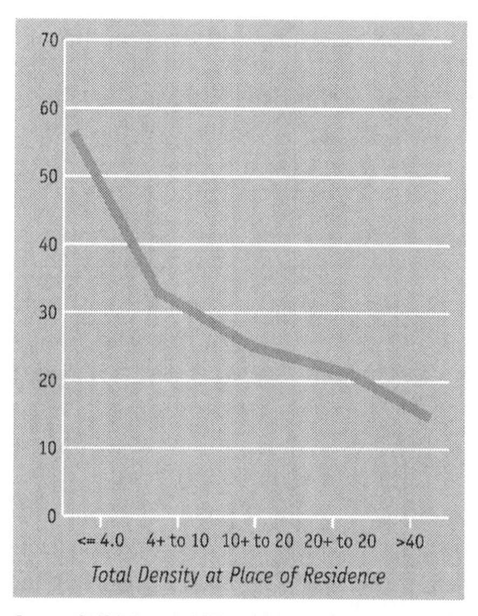

Figure 6.14 Focus on Density: Vehicle Miles Traveled per Household.
(Sacramento Area Council of Governments, 2008, p. 102).

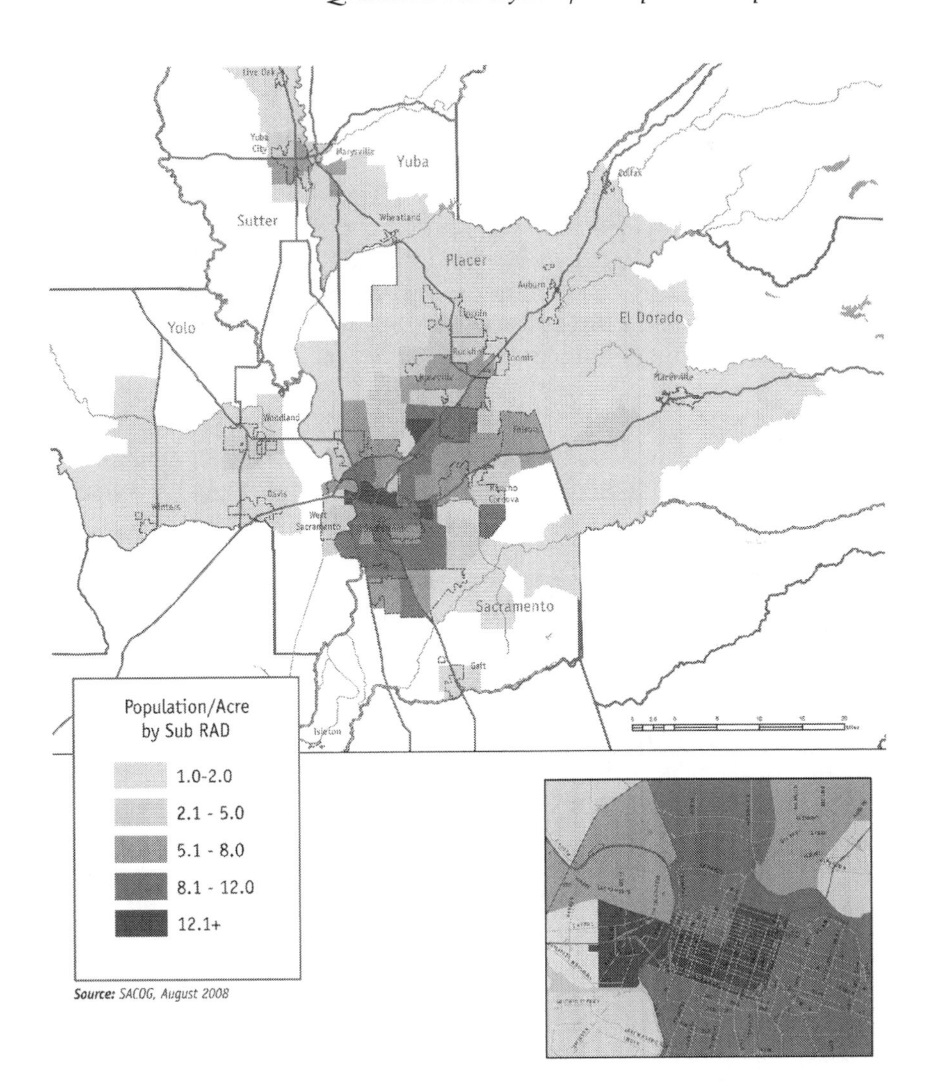

Figure 6.15 Focus on Density: Population per Acre.
(Sacramento Area Council of Governments, 2008, p. 97).

Or from the Cincinnati RTP:

> Higher densities in growing and infill areas can make transit more feasible by creating destinations and concentrated populations that may choose to use transit as an alternative to single-occupant automobile trips.
>
> (Ohio-Kentucky-Indiana Council of Governments, 2012, p. 4–4)

Whereas density is important, it appears to be the least important of D variables from a travel standpoint.

Best practice

Destination accessibility can be achieved by channeling growth into population and employment centers. Many MPOs now have center-based future land use plans for their regions. They seek to channel growth into these centers by channeling transportation investments into these centers, and providing discretionary funding to localities that accept and cultivate dense, mixed-use centers with interconnected streets (see section on Smart Growth).

Among the regions with center-based plans are Austin, Denver, Eugene, Minneapolis-St. Paul, Philadelphia, Portland, Salt Lake City, and Seattle (Capital Area Metropolitan Planning Organization of Austin, 2010; Denver Regional Council of Governments, 2011; Central Lane Metropolitan Planning Organization, 2011; Minneapolis-St. Paul Metropolitan Council, 2010; Portland Metro Regional Government, 2014; Wasatch Front Regional Council, 2015; Puget Sound Regional Council, 2010). The Austin RTP notes:

> Historically, CAMPO has developed long-range transportation plans based on past growth trends. We have taken a different approach for the current plan. This plan has been developed with the assumption that we may no longer be able to afford to invest in major regional infrastructure as we have in the past . . . . The CAMPO 2035 Plan assumes that the region will work toward implementation of a network of high density mixed use centers oriented around the transportation investments included in the plan.
>
> (Capital Area Metropolitan Planning Organization of Austin, 2010, p. 17)

The Austin MPO is supporting the emergence of mixed-use activity centers throughout the region by setting aside funding for transportation projects serving centers identified on the CAMPO Centers Map.

Likewise, in Seattle:

> Transportation 2040 supports development of centers throughout the region. Centers are locations with compact, pedestrian-oriented development and a mix of different commercial, civic, entertainment and residential uses. While relatively small geographically, centers are strategic places identified to receive a significant proportion of future population and employment growth when compared to the rest of the urban area.

	CURRENT	2035 TARGET
Population Accommodation	16% of regional population within designated center	31% of regional population within designated centers
Employment Accommodation	36% of regional employment within designated centers	38% of regional employment within designated centers

Figure 6.16 Population and Employment Targets for Regional Centers.
(Capital Area Metropolitan Planning Organization of Austin, 2010, p. 22).

> Concentrating growth in centers allows cities and other urban service providers to maximize the use of existing infrastructure, make more efficient and less costly investments in new infrastructure, and minimize the environmental impact of urban growth.
>
> (Puget Sound Regional Council, 2010, p. 9)

A few plans (very few) also emphasize land use diversity (e.g., jobs-housing balance) and pedestrian-friendly design (e.g., local street connectivity).

Smart growth

Background

The term "smart growth" encompasses planning that proactively integrates land use and transportation through the establishment of multiple mixed-use urban centers, communities, and activity centers connected by regional transit and supported by local infrastructure networks for walking and biking.

A broad analysis of 85 regional land use-transportation scenarios showed that smart growth or compact growth scenarios reduced year 2050 VMT by 17 percent in comparison to trend land use-transportation development scenarios (Bartholomew & Ewing, 2008). This result offers important implications for regional transportation development, in part due to the need for immediate and massive reductions in VMT necessary to achieve climate change mitigation goals. These scenarios also help create economically,

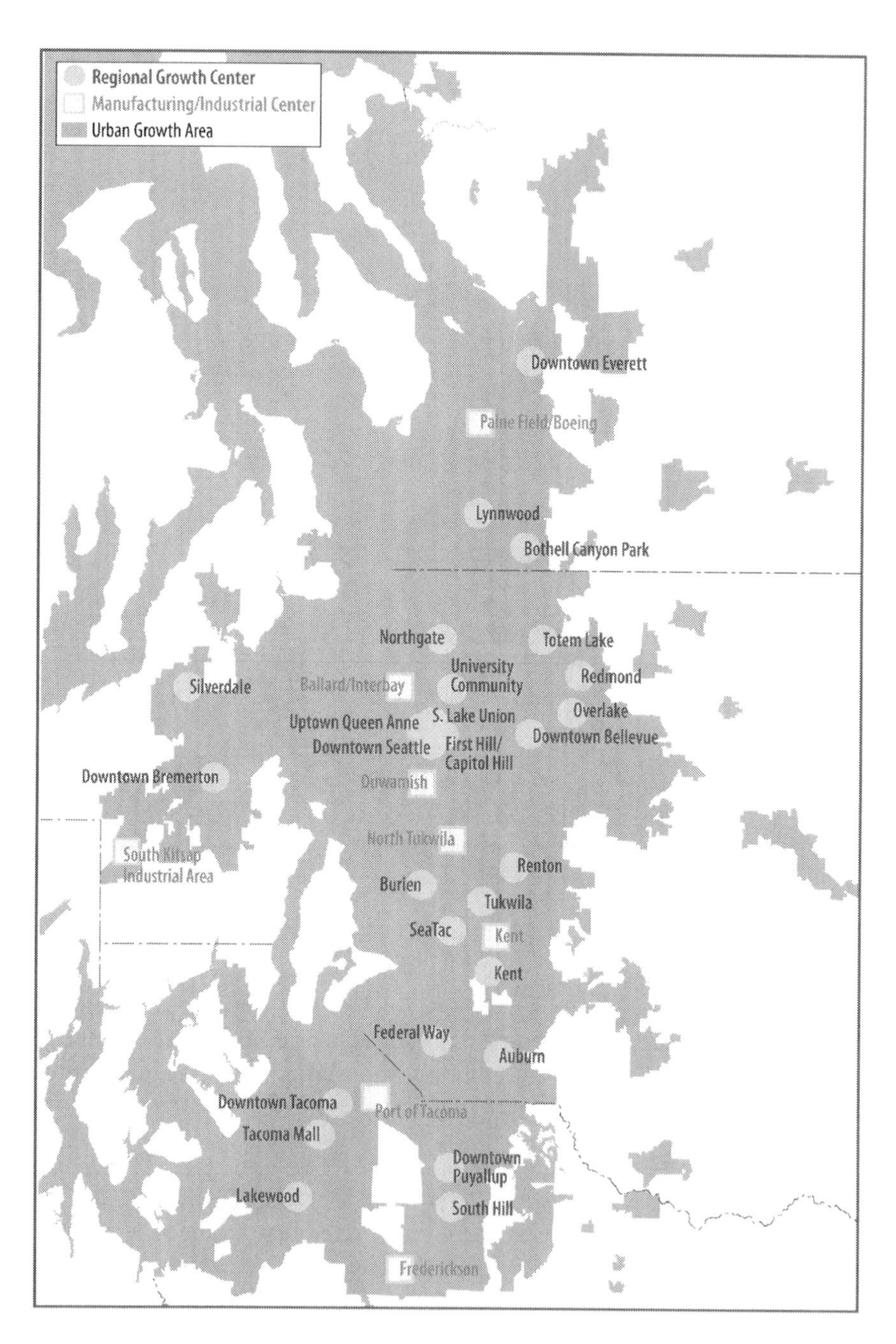

Figure 6.17 Center-Based Land Use Plan.
(Puget Sound Regional Council, 2010, p. 9).

environmentally, and socially relevant regional plans, in part through promotion of smart growth development that addresses travel behavior goals (Bartholomew & Ewing, 2008).

Conventional practice

The conventional practice is for MPOs to assume that development patterns are the exclusive purview of the development industry and local governments exercising zoning powers. This is changing, as indicated in the 2007 Boston RTP:

> It is important to coordinate transportation planning decisions and land use planning decisions so they are complementary, not contradictory. This is difficult, since transportation funding decisions are made at the regional and state levels, and land use decisions are primarily made by municipalities.
>
> However, extended public discussion on the relationship between land use, transportation, and economic development has clarified their links and has guided state, regional, and much local project-based decision-making in the direction of considering all three elements.
>
> (Boston Region Metropolitan Planning Organization, 2007, p. 11–2)

The practice is changing because the performance of RTPs depends as much on development patterns as on infrastructure investments. The 2007 Seattle RTP notes:

> While considerable attention is traditionally given to multi-modal infrastructure and transportation services investments, less attention is directed at regional development patterns and system management practices, such as the pricing of transportation and technological management systems, as a means to influence travel and reduce overall personal and public costs of regional transportation. Within the next 30 years, the region can meet its travel needs in a far more effective and cost efficient manner than it has during the past 30 years.
>
> (Puget Sound Regional Council, 2007, p. 16)

Best practice

Smart growth is now called for in many RTPs, reflecting state and local efforts to discourage sprawl.

What differs from region to region is not so much the desire for smart growth or the projected benefits, but the degree to which it is actively encouraged in the RTP.

Some MPOs seek voluntary compliance by local governments with smart growth plans, using fiscal and other analyses to show that smart growth

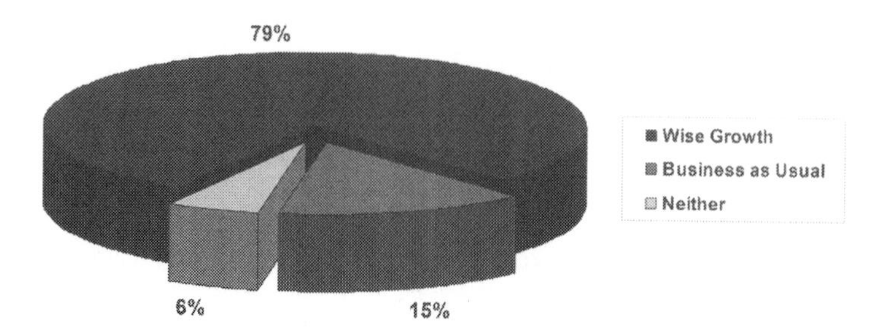

Figure 6.18 "Wise Growth" Favored and Adopted in Lansing, MI.
(Tri-County Regional Planning Commission, 2010, p. 2–60).

is more beneficial than sprawl. The Lansing MPO's "wise growth" plan is backed by such analyses. Through 2010, 44 governing bodies covering 80 percent of the region's population had adopted the future land use policy map and themes and principles, mostly by resolutions directing their staffs, consultants, and planning commissioners to implement that map and principles.

Still other MPOs have become even more proactive in incentivizing smart growth – by setting aside a portion of funds for local governments that will accept and encourage smart growth. This has emerged as a best practice. Because of its critical importance to VMT reduction, we describe the approaches used in three different regions.

San Francisco Bay

Under a program called FOCUS, 60 local governments in the San Francisco Bay region have designated Priority Development Areas (PDAs) within their jurisdictions. A PDA is locally designated land where future growth can be channeled, at sufficient densities to take advantage of existing infrastructure and services, especially transit service (Metropolitan Transportation Commission, 2009, p. 8).

FOCUS provides funding support via incentives such as capital infrastructure funds, planning grants, and technical assistance to these communities because they will bear the lion's share of the region's future growth. In this Transportation 2035 Plan, MTC doubles the size of its Transportation for Livable Communities program, to $2.2 billion over the next 25 years, in order to advance focused growth objectives and support PDAs. This, plus the $400 million set aside for the Transportation Climate Action Campaign, represent 1.19 percent of Bay Area's transportation revenues through 2035 (Metropolitan Transportation Council, 2009, pp. 46, 72).

The MPO estimates that people living in focused, compact neighborhoods of the type envisioned for PDAs travel 20 to 40 percent fewer vehicle miles each day than those who live in the sprawling suburban tracts that typify the Bay Area's post–World War II development pattern. This translates into a directly proportionate reduction in carbon dioxide emissions from personal travel (Metropolitan Transportation Council, 2009, p. 74).

The current inventory of adopted PDAs (planned and potential) includes nearly 120 individual areas across the region. Together they constitute only about 3 percent of the region's land area, but based on estimates provided by local governments, they could accommodate as much as 56 percent of the Bay Area's growth to the year 2035 – all in locations that will be accessible to high-quality transit. "The early interest in this program is a hopeful sign for the region . . . FOCUS Priority Development Areas (PDAs), in particular, serve as a mechanism to gain local government buy-in to pursue focused growth near transit nodes in their communities" (Metropolitan Transportation Council, 2009, pp. 8, 16).

San Diego

Under San Diego's 2005 Smart Growth Incentive pilot program, the MPO doled out $22.5 million in grants for local, smart growth development projects (San Diego Association of Governments, 2007, p. 1–4). Funding was from the federal Transportation Enhancements program. The success of the pilot program led to the creation of a $206 million Smart Growth Incentive Program funded by a half cent local sales tax. This is 2 percent of the revenues generated by the sales tax and 0.36 percent of the MPO's total revenues available through 2030 (San Diego Association of Governments, 2007, p. 5–26).

To be eligible for these funds, and receive priority for other funding, local governments designated almost 200 existing, planned, or potential "smart growth areas" to which dense mixed-use development is being directed. The areas fall into seven categories: Downtown San Diego, Urban Centers, Town Centers, Community Centers, Rural Villages, Mixed Use Transit Corridors, and Special Use Centers. These areas are shown on the accompanying Smart Growth Concept Map (see Figure 6.19). They currently house 22 percent of the region's population and are projected to house 29 percent by the year 2030. Every single jurisdiction in the San Diego region was able to identify at least one smart growth area on the map, demonstrating region-wide support for the smart growth principles included in the regional comprehensive plan (San Diego Association of Governments, 2007, p. 5–4).

The RTP establishes land use density and transportation service targets for each of the place types. If the areas on the map meet the targets in current land use and transportation plans, they are identified as "existing/planned" areas and eligible for infrastructure. If they do not, but they exhibit future opportunities for smart growth development, they are identified as "potential" areas eligible for planning grants only. Infrastructure grants can cover streetscape or sidewalk enhancements, transit station improvements,

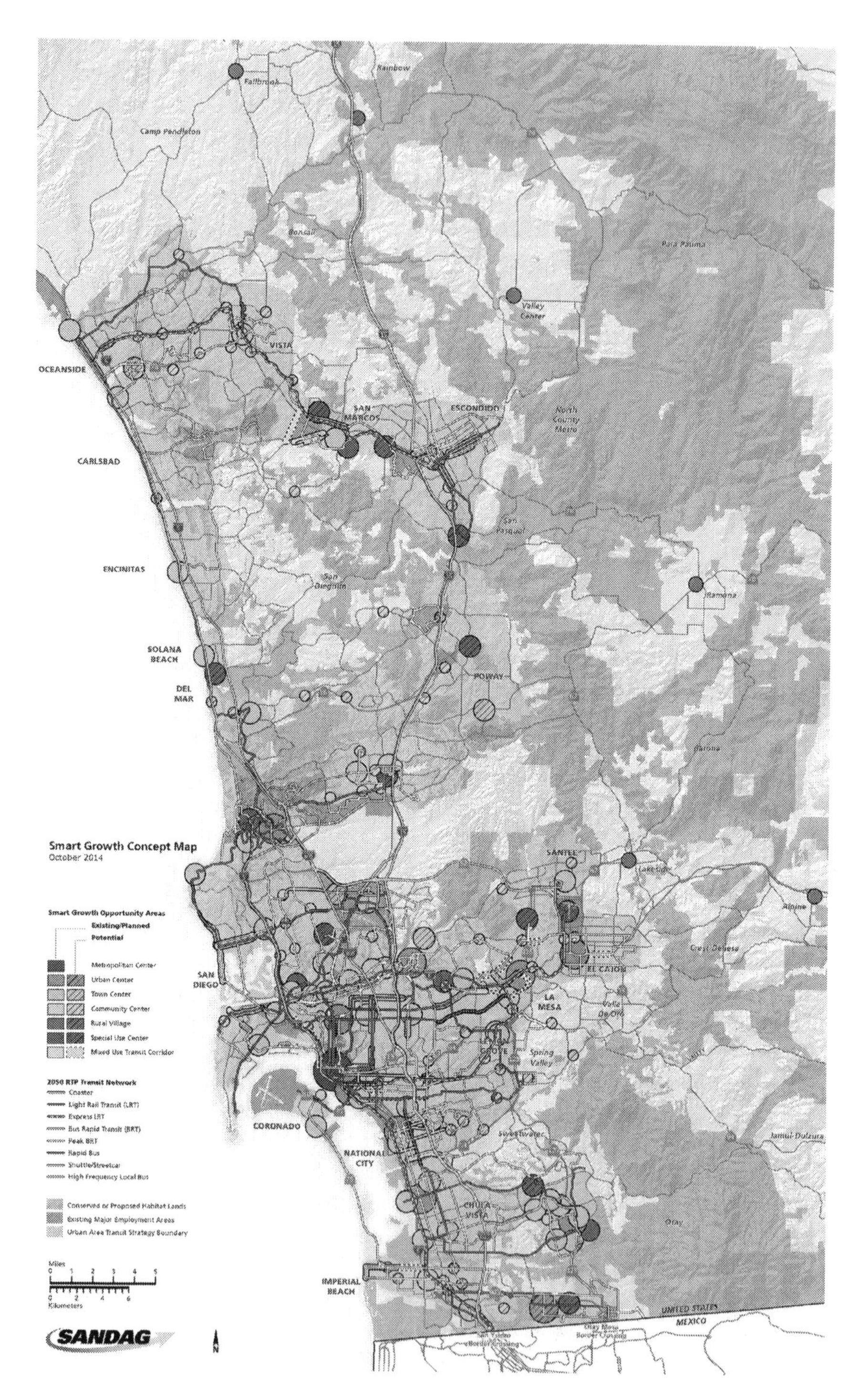

Figure 6.19 Smart Growth Concept Map.

(San Diego Association of Governments, 2007, p. 5–13)

traffic calming measures, or other quality of life amenities that support smart growth in that area. Planning grants can be used to amend general plans, prepare specific plans, or update zoning ordinances that will provide the institutional framework for smart growth development in these areas.

The smart growth area designations have been incorporated into the RTP Transportation Project Evaluation Criteria used to prioritize funding for regional transportation projects (San Diego Association of Governments, 2007, p. 5–3). The map guides the planning and development of the region's future transit networks, providing higher priority for peak period transit services that link smart growth areas to one another and to other major activity centers (San Diego Association of Governments, 2007, p. 5–16).

Dallas-Ft. Worth

Lest these examples be dismissed because they come from that outlier, California, the Dallas-Ft. Worth MPO has issued three Sustainable Development Calls for Projects (North Central Texas Council of Governments, 2013, p. D.31). To be eligible, projects have to be located in "Sustainable Development Areas of Interest," (see Figure 6.20) that is, within walking distance of current or potential future rail station locations; in areas with a concentration of unemployed persons, high-emitting vehicles, or low-income households; and/or in historic downtowns with multiple contiguous street block frontage of pedestrian-oriented development. Eligible projects are also required to have zoning in place that allows the project to be built by right.

Under this program, over $120 million have been awarded since 2001 (North Central Texas Council of Governments, 2013, p. 5.29). NCTCOG first call for projects funded 19 infrastructure projects and one planning project. The second call in 2005–2006 funded 29 infrastructure, four land banking, and nine planning projects. The 2009–2010 Sustainable Development Call for Projects funded 27 infrastructure projects and nine planning projects (North Central Texas Council of Governments, 2013, p. D.31). The RTP funds the program at a level of $625.6 million.

> Beyond earmarking funds for projects in smart growth areas, MPOs can encourage smart growth by providing technical assistance to localities that may not have the necessary expertise in-house. The most ambitious program to our knowledge is that of the Southern California Association of Governments' Compass Blueprint Program. The MPO's consulting services are available to all local governments in the region, free of charge.
>
> Since 2004, Compass Blueprint has been a model for integrating land use and transportation planning and turning regional vision into local reality. Guided by four core principles, Mobility, Livability, Prosperity and Sustainability, these efforts have effectively given the region a "jump-start" in implementing this SCS (Sustainable Communities Stategy). At

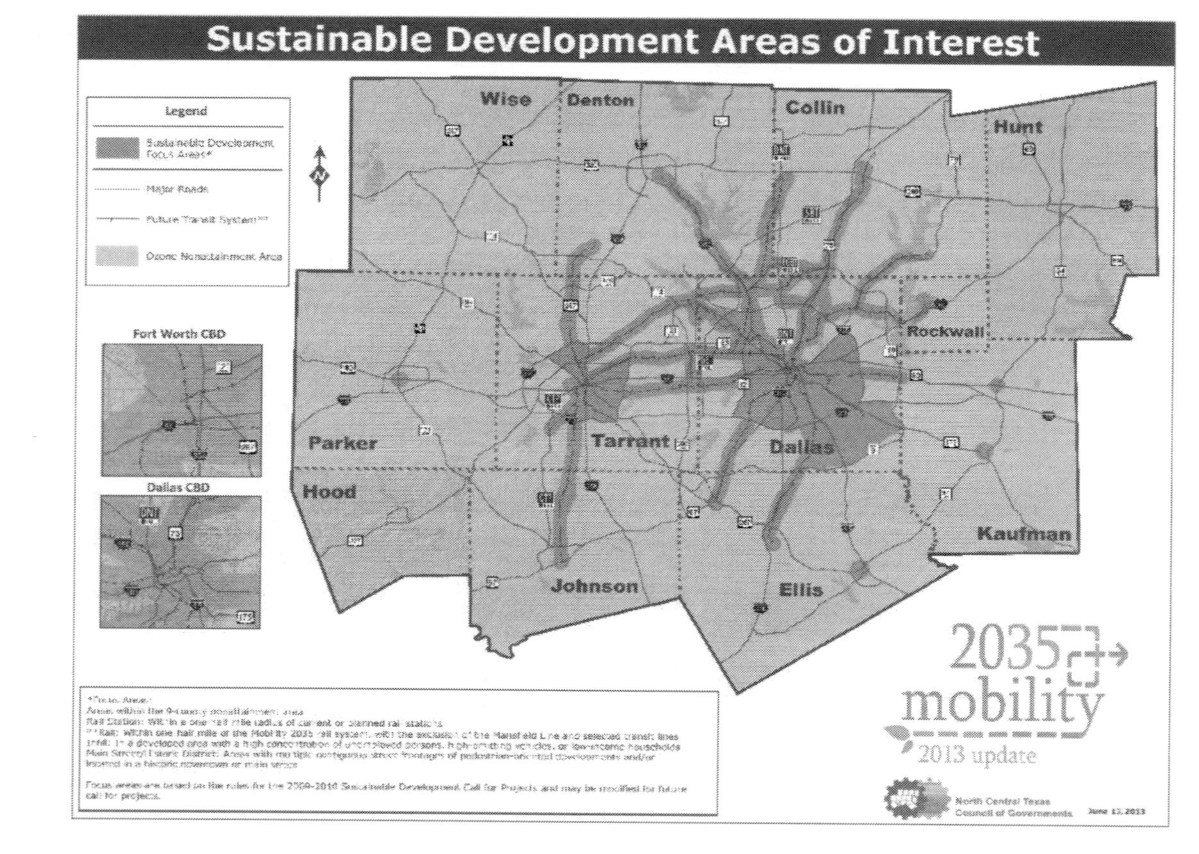

Figure 6.20 Sustainable Development Focus Areas in Dark Shading.
(North Central Texas Council of Governments, 2009, p. 5–30)

the core of Compass Blueprint are Demonstration Projects (see also Figure 6.21) – incentive-based, voluntary partnerships between SCAG and local governments that apply innovative approaches and tools to local plans that support regional priorities. As of September 2011, SCAG has provided over $10.5 million in incentive funds for 132 Demonstration Projects in 95 local jurisdictions. Projects have included transit-oriented development plans for station areas along new light-rail alignments, downtown revitalization efforts, community visioning projects in low-income communities, and other projects that support shared local and regional goals.

(Southern California Association of Governments, 2012, p. 120)

The provision of technical assistance to localities that may not have the necessary expertise in-house is not just the domain of large MPOs. The Wasatch Front Regional Council in Salt Lake City also has a technical assistance program known as the Transportation and Land Use Connection (TLC) that supports the region's vision, Wasatch Choice 2040.

The program helps communities implement changes to the built environment that reduce traffic on roads and enable more people to easily walk, bike, and use transit. This approach is consistent with the Wasatch

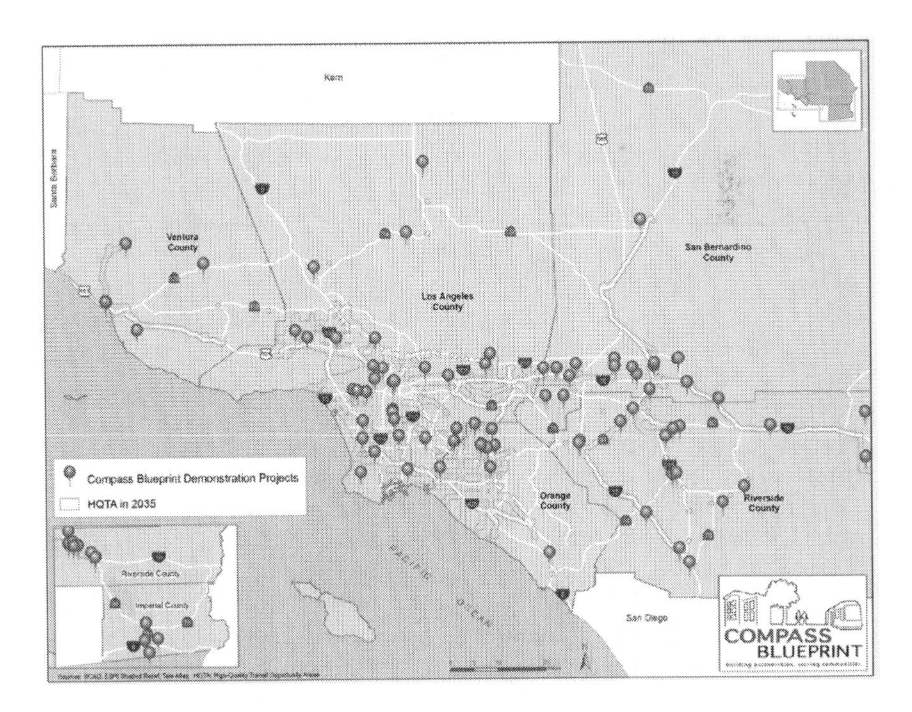

Figure 6.21 Compass Blueprint Demonstration Projects.
(Southern California Association of Governments, 2012, p. 121)

Choice Vision and helps residents living throughout the region enjoy a high quality of life through enhanced mobility, better air quality, and improved economic opportunities.

(Wasatch Front Regional Council, 2017, p. 1)

Over four years from 2014 to 2017, TLC assisted communities with 54 planning projects with a total project valuation of $4.38 million.

(Wasatch Front Regional Council, 2017, p. 2)

Major highway expansion

Background

Many regions address growing traffic congestion by trying to expand highways at the same pace as traffic grows. They aren't successful. All but one of the 101 urbanized areas for which data are available show substantially higher levels of congestion in 2014 than 1982, the first year for which data are available. This is despite trillions of dollars in highway expansion. Even Houston, which temporarily reduced congestion for years through a continuous program of highway construction, eventually succumbed to increasing traffic.

Substantial evidence shows that roadway expansion, in all likelihood, encourages vehicular traffic (see Chapter 5). In a series of studies on the Sacramento region, Johnston (2006) found that expanding roadways increases auto travel compared to doing nothing. Other evaluations find a clear association between increasing lane miles and increasing vehicle miles traveled. Noland (2001) found that 50 percent of added capacity fills in five years, and up to 80 percent of added capacity fills over the long run. Moreover, utilizing a cross-sectional time series model, Noland (2001) forecast that 25 percent of future VMT growth will be due to lane mile additions, assuming historical rates of growth in road capacity. A robust model by Cervero (2003) assessed induced travel demand and induced growth in California. Outcomes showed that while not significant in all settings, roadway investments clearly influence travel demand, with induced travel growing over time.

In assessments comparing transportation investments, highway expansion consistently shows the lowest potential return. A 2001 assessment of potential investments in Cincinnati, Ohio, explored options including an expanded regional bus service, light rail, and highway expansion. In this analysis, even large investments in highway expansion produced very little economic improvement (HLB Decision Economics, 2001). A more recent study for Atlanta, Georgia, found that better management of current transportation options provided substantially greater returns when compared with new highway construction (McKinsey and Company, 2008).

One way to potentially blunt the VMT-inducing effect of additional highway capacity and roadway expansion is to combine highway expansion with user pricing and high-occupancy vehicle (HOV) treatments. HOV facilities provide preferential treatment for transit, vanpools, carpools, and other vehicles by providing lanes and roadways reserved for their use. Travel time savings and reliability improvements underlie the attractiveness of HOV facilities for users. These benefits may accrue from short queue bypass HOV lanes as well as longer facilities, particularly at bottlenecks along high-type facilities, or at geographic barriers like water crossings.

Whereas these strategies will not reduce, in absolute terms, regional traffic congestion, a diversity of modes may more effectively moderate increases in congestion by accommodating new person trips. Combined with innovative management strategies and regional land use-transportation coordination, diverse multimodal investments offer a greater return on investment than traditional highway expansion. Roadway expansion in some form may be part of an investment package; however, the negative effects of induced travel and induced development must be weighed against the positive, at least temporary, congestion relief provided by major highway expansion projects.

Conventional practice

The conventional practice is to expand highway capacity through general purpose lane additions. Such expansion induces additional traffic and urban sprawl. If pricing is applied, it is in the form of general tolls that do not vary by time of day or congestion level. If lanes are reserved for high-occupancy vehicles, the lanes are not tolled and hence do not generate needed revenues. An example is the set of highway improvements that make up the Chicago RTP (Chicago Metropolitan Agency for Planning, 2008). There are sprawl-inducing general lane additions such as those on I-55 and I-57 in the southern exurbs. There are toll road expansions such as those on I-88 and I-90 in the western exurbs. And there is a conventional (not tolled) HOV expansion on I-290 (see Figures 6.22–6.24).

Best practice

Federal law and regulation already strive to discourage general purpose highway expansion in ozone and carbon monoxide nonattainment areas, acknowledging the induced traffic generated by such expansion:

> (e) In TMAs designated as nonattainment for ozone or carbon monoxide, the congestion management process shall provide an appropriate analysis of reasonable (including multimodal) travel demand reduction and operational management strategies for the corridor in which a project that will result in a significant increase in capacity for SOVs (single-occupant vehicles) is proposed to be advanced with Federal funds. If the analysis demonstrates that travel demand reduction and operational management strategies cannot fully satisfy the need for

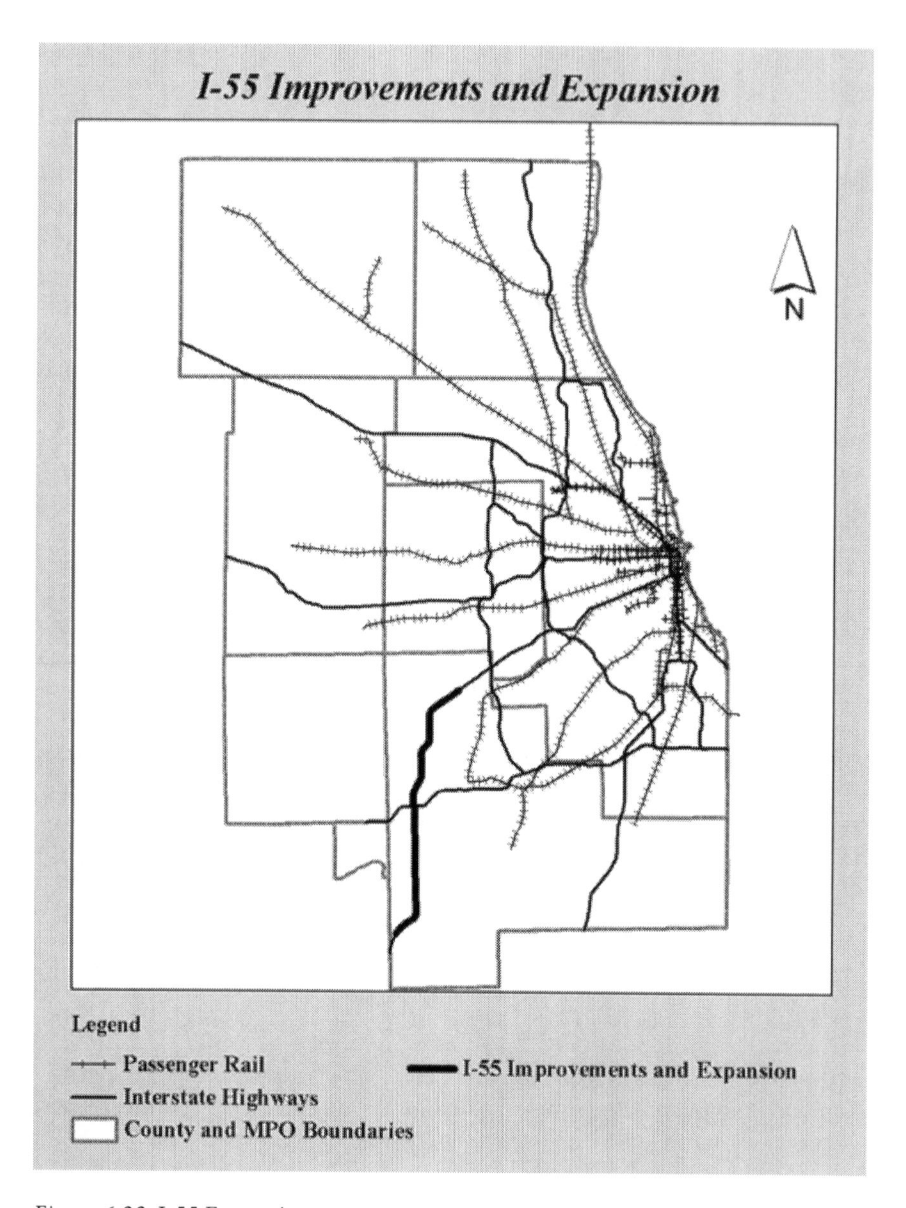

Figure 6.22 I-55 Expansion.
(Chicago Metropolitan Agency for Planning, 2008, p. 192)

additional capacity in the corridor and additional SOV capacity is warranted, then the congestion management process shall identify all reasonable strategies to manage the SOV facility safely and effectively (or to facilitate its management in the future). Other travel demand reduction and operational management strategies appropriate for the corridor, but

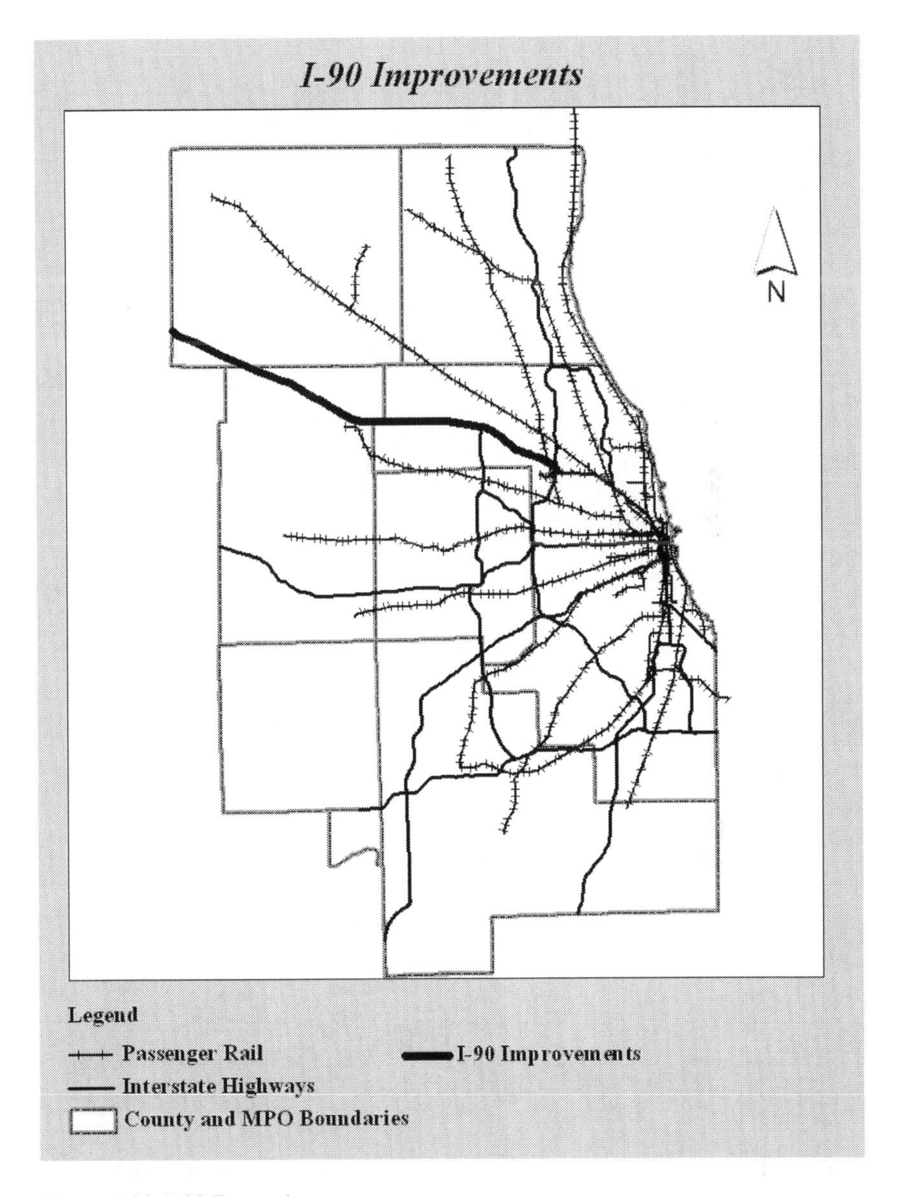

Figure 6.23 I-90 Expansion.

(Chicago Metropolitan Agency for Planning, 2008, p. 178)

> not appropriate for incorporation into the SOV facility itself, shall also be identified through the congestion management process. All identified reasonable travel demand reduction and operational management strategies shall be incorporated into the SOV project or committed to by the State and MPO for implementation.
>
> (23 CFR 450.320(e))

Figure 6.24 I-290 Expansion.
(Chicago Metropolitan Agency for Planning, 2008, p. 197)

An emerging best practice is to favor managed facilities over general purpose highway expansion. Although commonly employed by airlines, utility companies, and others, using price to avoid peak-period overload is the exception in surface transportation policy. As demonstrated by successful implementation in many U.S. cities, high-occupancy toll (HOT) lanes – which allow noncarpool drivers to pay a toll to access underutilized carpool lanes – can bring real benefits to travelers. HOT lanes, often called express lanes, provide travel options for carpools, express buses, and toll payers; they allow for more efficient use of freeway capacity; and they generate revenues for other highway and transit improvements (see Figure 6.25).

Express lanes have been in operation for more than 15 years in Houston, Los Angeles, and San Diego, and in the past 10 years have opened in Denver, Miami, Minneapolis, Salt Lake City, and Seattle, among others. Other regions are considering installation of express lanes. "An integrated system of Express Toll Lanes could help ease the impact of traffic congestion on Marylanders' lives and do so decades sooner than traditional approaches would allow" (Baltimore Regional Transportation Board, 2007, p. 151).

Surveys show most travelers use express lanes to bypass congestion when they are late to pick up a child at daycare, to squeeze more working hours out of a day, or to catch a plane. For this reason, and because revenue from express lanes often supports public transit service enhancements, express lanes are widely supported by travelers at all income levels (Metropolitan Transportation Commission, 2009, p. 19).

The San Francisco 2035 RTP establishes and funds the Bay Area Express Lane Network. The initial segment of the Network opened in 2010 on a 14-mile stretch of Interstate 680 between Pleasanton and Fremont. Additional express lanes opened in 2011 on I-580 through the Tri-Valley and on SR 237/I-880 Connector in Santa Clara County. The 2035 RTP would extend the express lane concept to a connected network of express lanes spanning 800 miles, greatly improving travel options and freeway efficiency throughout the Bay Area (Metropolitan Transportation Commission, 2009, p. 60).

The San Francisco MPO estimates that it will cost $7.6 billion to build, finance, and operate the Bay Area Express Lane Network over the next 25 years. With gross express lane toll revenues reaching $13.7 billion over the same period, the remaining $6.1 billion in net revenue would be available to finance additional improvements in the express lane corridors.

The Bay Area Express Lane Network is a strategy to accelerate completion of the region's carpool and public transit system. MTC will convert to express lanes some 400 miles of carpool lanes that already exist or are under construction, plus 100 new miles of fully funded lanes will be built in the next four years. The revenue generated will also be used to construct some 300 new miles of express lanes that close gaps and extend the system. In total, the 300 new miles amount to less than a 6 percent increase in total Bay Area freeway mileage, and more than half the added mileage is for gap

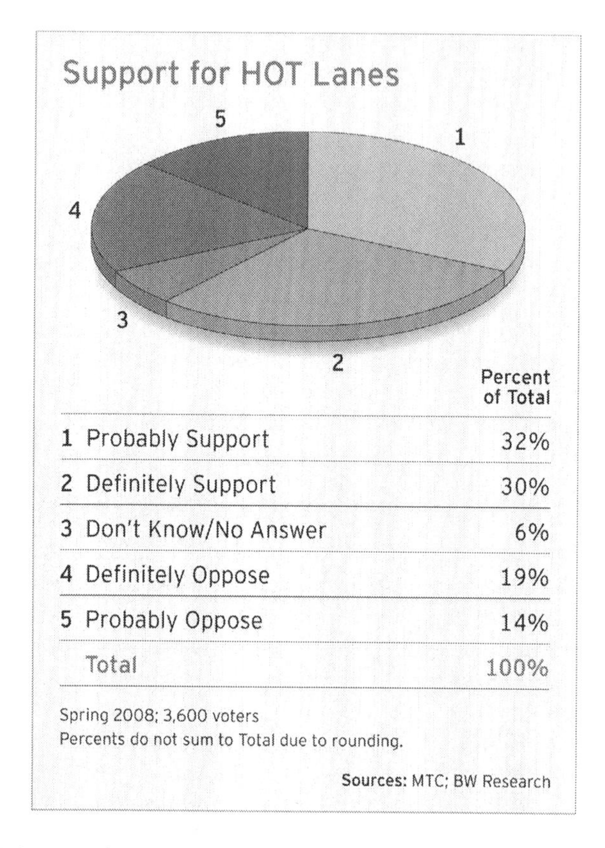

Figure 6.25 Support for HOT Lanes in the Bay Area.
(Metropolitan Transportation Commission, 2009, p. 19).

closures that connect two existing carpool lanes (Metropolitan Transportation Commission, 2009, p. 64).

To keep express lane traffic flowing freely, toll rates adjust dynamically to balance supply and demand based on data from roadway sensors used to monitor traffic conditions. Tolls during the most congested periods, when carpool and bus traffic is heavy, are comparatively high so only a small number of noncarpoolers – those who most need congestion insurance that day – buy in. Tolls are much lower during periods of lighter traffic. Noncarpoolers using the express lanes pay their tolls through the FasTrak® system already in place on the region's eight toll bridges. With FasTrak® readers installed on overhead structures, tolls can be collected without forcing drivers to slow down or stop (Metropolitan Transportation Commission, 2009, p. 61).

San Diego also has plans for an extensive network of HOT lanes on Interstates 5, 15, and 805 and HOV facilities on State Routes 52, 78, 94, and 125 (totaling more than 200 mainline miles). The managed lane facilities on

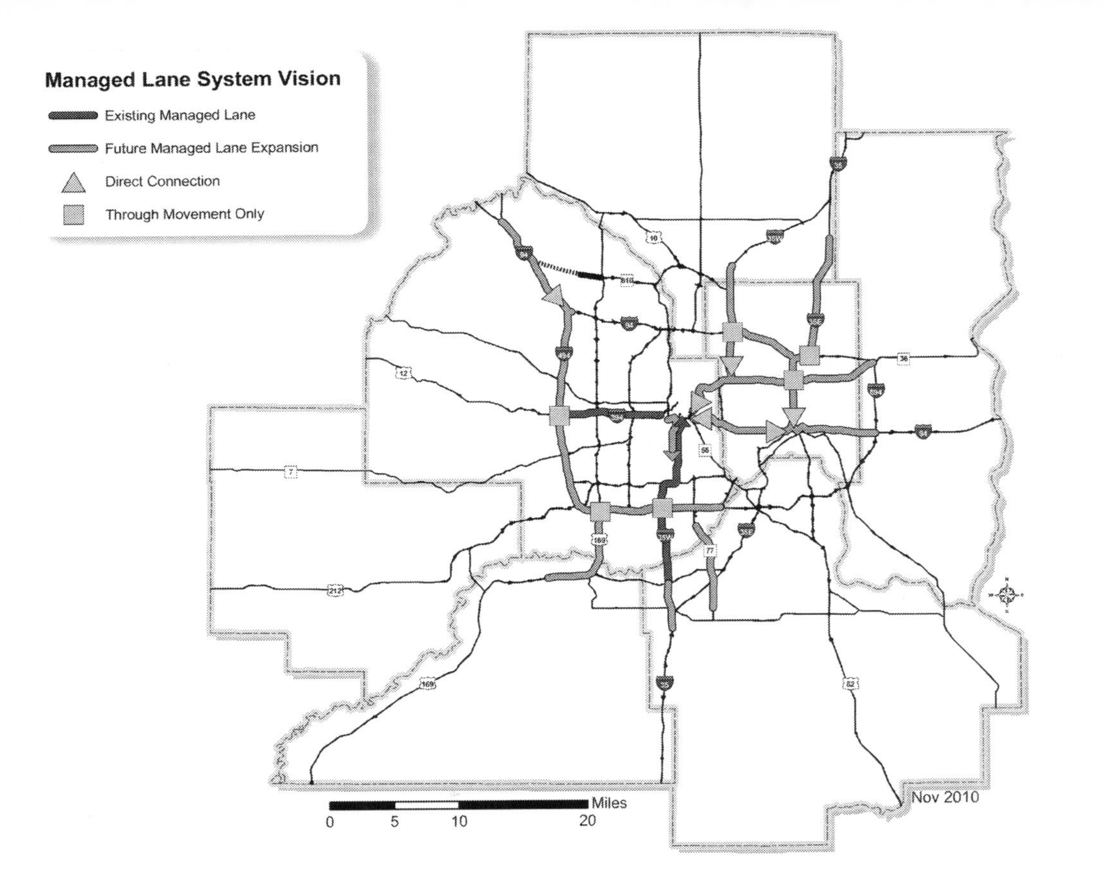

Figure 6.26 Existing and Planned Managed Lane Facilities.

(Published with permission of the Metropolitan Council, 2030 Transportation Policy Plan, adopted by the Council in November 2010, Chapter 6 (Highways) Minneapolis-St. Paul Metropolitan Council, 2010, p. 97).

Interstates 5 and 805 are modeled after the I-15 Managed Lanes project, the first in the nation. Pricing is used to cover selected connectors and a network of ramp meters in the region (San Diego Association of Governments, 2007, pp. 7–14, 7–15).

Other regions with less ambitious plans to institute or expand express lanes include Atlanta, Denver, Los Angeles, Minneapolis-St. Paul (see Figure 6.26), and Seattle.

Minor street expansion

Background

As a subject, the design of minor street networks has slipped through the cracks between engineering and planning. Yet, there is evidence that travel and traffic are as sensitive to street networks as to land use patterns (Ewing et al., 2018).

To disperse traffic and avoid bottlenecks, the street network must provide many connections between activities. To encourage walking, bicycling, and transit use, the network must provide reasonably direct routing. Curvy, discontinuous streets are no problem for automobiles at 35 miles an hour, but make trips too long for pedestrians at 3 miles an hour. Such streets force buses to backtrack and take indirect routes, and make it harder to provide service within easy walking distance of residents.

Ewing and Cervero (2010) conducted a meta-analysis of more than 200 studies of the built environment and travel, identifying generalized elasticities of travel with respect to change in measures of built environment characteristics. This analysis showed that street design and network connectivity are primary determinants of VMT, walking, and transit use. In particular, intersection/street density and the proportion of four-way intersections were two of the most important negative influences on vehicle miles traveled. Thus, a more complete local street network translates into reduced VMT. These street design measures also represented two of the most important positive influences on transit use and walking. The elasticities computed by Ewing and Cervero (2010) suggest that a 10 percent increase in intersection density will cause a 1.2 percent reduction in VMT, a 3.9 percent increase in walking, and a 2.3 percent increase in transit use.

A review of regulations in 14 U.S. communities provides examples of requirements aimed at all three network characteristics: (1) block size, (2) degree of curvature, and (3) degree of interconnectivity (Butler, Handy, & Paterson, 2003). With block length limits, land development codes control the spacing of streets, thereby creating relatively predictable and evenly distributed networks of streets. With connectivity requirements, land development codes require a certain ratio of street segments to street ends, thereby effectively limiting the number of cul-de-sacs and loop roads. While much less common, the ratio of travel distance via the network to straight-line distance between points is also sometimes used to affect block size, curvature, and connectivity. Block length limits vary from 200 to 1,500 feet. Some ordinances, such as the ordinance for Charlotte, NC, set different limits for streets at different levels in the functional hierarchy. This form of access management is becomingly increasingly common as localities strive to maintain through-capacity on their highest-order roads.

In one notable example of minor street regulation on a broad scale, the state of Virginia maintains secondary street acceptance requirements. Virginia is one of a few states that manage the maintenance and operations of local streets, providing an opportunity for direct influence on street design. These secondary street requirements emerged after Virginia began accepting maintenance responsibilities for more streets at a time of increasing congestion and declining revenues. Through these standards, the state seeks to achieve these goals, as well as broader aims, including reducing future construction needs and operational costs, improving emergency response times, improving safety and access for pedestrians and bicyclists, creating a more efficient transportation network, encouraging appropriate speeds, and reducing stormwater runoff. Major concepts of the regulation include adding streets to the state system as a network rather than as individual segments, incorporating context-sensitive design (with narrow lanes and flexible parking standards), and providing a variety of designs according to area type (compact, suburban, and rural, as defined by zoning regulations). Before the Virginia Department of Transportation will accept responsibility for maintaining local streets, there are certain connectivity requirements that must be met. The connectivity requirements aim to link adjacent developments and undeveloped parcels, thus improving the overall capacity of the transportation network, reducing vehicle miles traveled, and improving emergency response times.

Conventional practice

The conventional practice is to focus attention and funding on freeways and principal arterials. If they are projected to operate below some acceptable level of service, then the RTP includes road widening or operational improvements among the funded projects. Typical of this emphasis are the following:

> Completing the Regional Arterial System is a priority in the 2050 RTP. Regional arterials provide critical links to the highway network, and they serve as alternative routes to highways as well.
>
> (San Diego Association of Governments, 2011, p. 6–39)

> Roadway capacity expansion projects include over 1,500 new miles of highway and regional arterial lanes to address the region's worst choke points, complete projects that have been started, and anticipate future problems. This represents a 13 percent increase in regional arterial and state freeway system lane miles.
>
> (Puget Sound Regional Council, 2007, p. ii)

> The 2040 RTP includes new or widened arterial streets and freeway improvements identified as needed to serve the existing and developing areas of the Wasatch Front Region. Approximately 1,071 lane miles of capacity improvements are planned for the next 30 years.
>
> (Wasatch Front Regional Council, 2011, p. 154)

The problem with this focus is twofold. First, it ignores the important role played by lesser roadways in the functional hierarchy, those roads that are classified as minor arterials, collectors, or local streets. The literature suggests that a more complete local street network can actually reduce VMT by shortening travel distances (Ewing and Cervero, 2010).

Another problem with the focus on freeways and principal arterials is the incontrovertible fact that improvements to these higher-level facilities induce additional VMT. This is apparent from the literature generally (Cervero, 2003; Ewing et al., 2008) as well as quantitative analyses in Chapter 5.

Best practice

Depending on how complete the network is, minor arterials, collectors, and local streets can either relieve some of the pressure on arterials, or add to that pressure. Also, depending on the network, minor streets can make walking and cycling either attractive or nearly impossible. A few RTPs acknowledge the importance of having a dense network of interconnected minor streets:

> The curvilinear cul-de-sac street pattern typical of recent subdivision design in the OKI region usually has very long blocks and many dead end streets. This pattern offers few route options since all traffic is typically funneled out onto a small number of arterial roads, causing congestion. Connectivity involves a system of streets providing multiple routes and connections to the same origins and destinations. Improving street connectivity by providing parallel routes and cross connections, and a small number of closed end streets can reduce traffic on arterial streets and reduce travel time.
>
> (OKI Council of Governments, 2012, p. 4–4)

> A better-connected network of neighborhood streets will help relieve traffic growth along heavily used corridors, and reduce congestion at major choke points and intersections. These streets will also provide for many safety improvements to the overall transportation network, allowing people to access nearby destinations on smaller-scaled, walkable, bikeable, and transit-friendly roadways.
>
> (Charlottesville Albemarle Metropolitan Planning Organization, 2004, p. 4)

> Recent research has shown that block sizes within much of the urban growth area are often large and scaled to cars rather than pedestrians . . . Analysis of urban form and basic infrastructure in urban centers and clusters in the region suggests that deficiencies in these areas have significant performance impacts on both the local and metropolitan transportation systems.
>
> (Puget Sound Regional Council, 2007, p. 27)

Very few MPOs include local and developer funding for minor roads in their plans, budget regional resources for them, or establish street connectivity guidelines. The Denver RTP budgets for them:

> While local streets are not depicted as part of the regional roadway system, they are important for providing access to and through local developments and neighborhoods. The costs to build and maintain local streets, including collectors and minor arterials, are included in the 2035 MVRTP.
>
> (Denver Regional Council of Governments, 2011, p. 48)

The Sacramento RTP allocates regional resources to minor roads. By the mid-1970s, the Sacramento region had decided not to expand its freeway system further. Roadway investments in the 2035 RTP are instead of two principal types:

> $2.9 billion ($5.0 billion in escalated costs) goes to state highway improvements, mainly to complete four-lane highways to connect the northern counties with the rest of the region and add carpool lanes to urban freeways.
>
> $8.3 billion ($14.3 billion in escalated costs) goes to local road improvements, including intersection improvements, safety projects, signal timing, road widening in growth areas, and new connections for local access.
>
> (Sacramento Area Council of Governments, 2008, p. 9)

The City of Charlotte has established a connectivity policy that emphasizes a system of streets providing multiple routes and connections between origins and destinations.

> Connectivity is important because a highly connected street network can greatly reduce trip lengths, thereby reducing vehicle miles travel which in turn results in reduced emissions.
>
> (Mecklenburg-Union Metropolitan Planning Organization, 2010, p. 8)

Highway preservation and maintenance

Background

It is important to avoid policy and planning practices that favor investments in new transportation facilities if existing facilities are inadequately maintained and managed. A management philosophy called "asset management" or "fix-it-first" emphasizes the importance of maintaining and optimizing use of existing facilities.

Between 2009 and 2011, the latest year with available data, states collectively spent $20.4 billion annually to build new roadways and add lanes to existing roads. America's state-owned road network grew by 8,822 lane-miles of road during that time, accounting for less than 1 percent of the total in 2011 (Smart Growth America, 2014).

During that same time, states spent just $16.5 billion annually repairing and preserving the other 99 percent of the system, even while roads across the country were deteriorating. On a scale of good, fair, or poor, 21 percent of America's roads were in poor condition in 2011. Just 37 percent of roads were in good condition that year – down from 41 percent in 2008 (Smart Growth America, 2014).

The American Society of Civil Engineers publishes an annual Report Card for America's Infrastructure, covering transportation, water and environment, public facilities, and energy. ASCE gave the nation's roadways as a whole a near-failing D grade in 2013. Only bridges were graded higher, receiving a grade of C+. The grades may be self-serving to the construction industry and to the civil engineering profession, but that serious problems exist is not in dispute. ASCE estimates that poor road conditions cost motorists $67 billion a year in repairs and operating costs, and cost 14,000 Americans their lives (American Society of Civil Engineers, 2010).

Exacerbating such prices, deferred roadway maintenance often increases costs beyond that indicated by actual expenditures, as spending fails to keep pace with deteriorating conditions (Litman, 2009b). Clearly, failing to adequately fund pavement preservation in the short term creates substantial long-term costs, as visualized in Figure 6.27. Every $1 spent on preservation can eliminate or delay the need for $6 to $14 for road reconstruction. Thus, regions with many years of deferred maintenance projects face significant costs in the coming decade.

Return on maintenance investment varies according to the targeted roadway, but given that the majority of highway travel now takes place on roadways below "good" condition, there is no shortage of opportunities to make high return on investment (ROI) expenditures on repair and maintenance (Schrank & Lomax, 2009). An international study found that the return on road maintenance projects financed by the World Bank over the period 1961 to 1988 was about 38.6 percent compared to about 26 percent for all

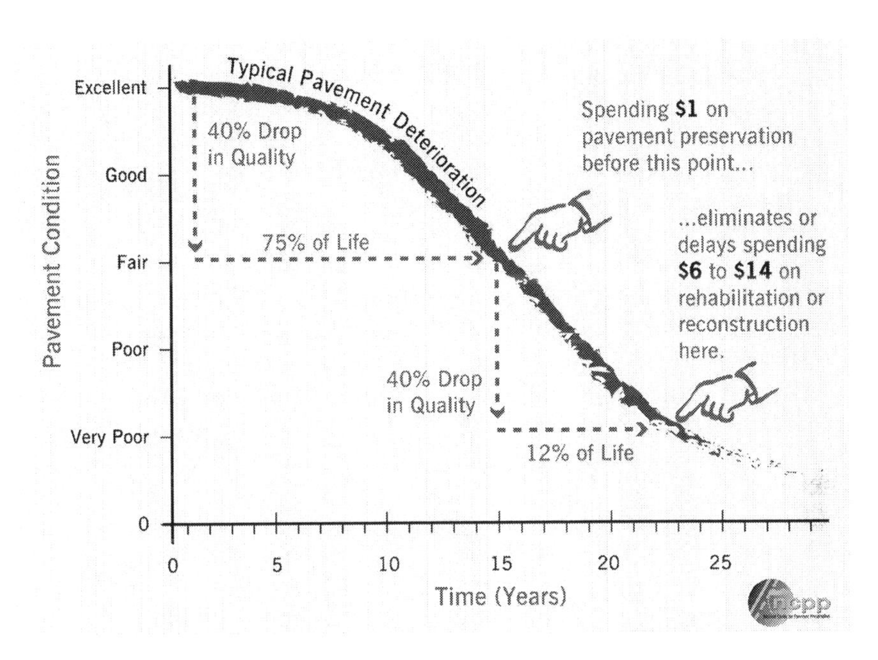

Figure 6.27 Allowing Pavement to Deteriorate Increases Costs.
(Madsen, Davis, & Baxandall, 2010, p. 17).

transportation projects, and 21 percent for all investment projects generally (Heggie & Vickers, 1998). Current analysis confirms a similar outcome for the United States (Kalyvitis and Vella, 2011). In sum, maintenance and repair provide the highest possible ROI of any widely available investment.

Conventional practice

The conventional practice is to split available revenues between highway maintenance and highway expansion, leaving both about equally underfunded relative to the amounts determined in needs assessments. The Sacramento RTP notes:

> Road programs are so significantly underfunded today that funding for road expansion must compete with funding for road maintenance, rehabilitation, and operations. The cycle typically unfolds as follows: Some road maintenance must be deferred; the road deteriorates to the point it must be reconstructed, which costs more but becomes eligible to use capital funds; so capital funds are siphoned off for road repair.
>
> (Sacramento Area Council of Governments, 2012, p. 239)

Similarly, the San Diego RTP notes:

> As the region's transportation system ages, the ongoing costs to maintain our existing infrastructure require a significant share of our future transportation funds . . . The limited revenues under the other [reasonable] scenarios were not sufficient to fund the full level of estimated highway rehabilitation needs.
>
> (San Diego Association of Governments, 2007, p. 4–12)

In the 2011 RTP, road maintenance and rehabilitation needs exceed existing revenues by 6.4 billion (San Diego Association of Governments, 2011, p. 5–4).

The Denver RTP has $23.4 billion to preserve and maintain the regional roadway system, far less than the $35.1 billion needed (Denver Regional Council of Governments, 2011, p. 1). The Portland RTP refers to a $422 million backlog of unmet maintenance needs in the City of Portland. "Without new revenue, that backlog is expected to continue growing at a rate of $9 million per year" (Portland Metro Regional Government, 2010, p. 1–26). The Austin RTP notes that maintenance has been deferred on the Texas state highway system. "Recently, the Texas Transportation Commission has recommended shifting future resources from added capacity projects to maintenance activities to cover ballooning maintenance needs" (Capital Area Metropolitan Planning Organization of Austin, 2010, p. 62).

Interestingly, highway maintenance needs are left unfunded even in some regions where the stated priority is system preservation, or "fix it first." Los Angeles gives priority to "maintenance and preservation" over "system completion and expansion." Yet, the 2008 RTP reports a funding shortfall for highway maintenance and preservation of $30 billion through 2035 (Southern California Association of Governments, 2008, p. 17).

The Colorado Springs RTP states: "Neglecting the preservation needs of the Pikes Peak region is not a rational policy choice; deferring maintenance due to fiscal pressure would necessitate spending substantially more on transportation investments in the future" (Pikes Peak Area Council of Governments, 2012, p. 62). Highway maintenance was a top priority in both rounds of a hypothetical spending exercise in which public participants were asked how they would like to spend available funds. Yet, the Colorado Springs RTP leaves highway maintenance and operation underfunded by a wider margin than new construction.

> For the regional roadway system, costs to maintain existing lanes from 2011 through year 2035 exceed the available revenues by approximately $2 billion. The current backlog of needed roadway maintenance is in excess of $500 million, and the backlog of bridge and major infrastructure maintenance has been estimated at $400 million.
>
> (Pikes Peak Area Council of Governments, 2012, p. 130)

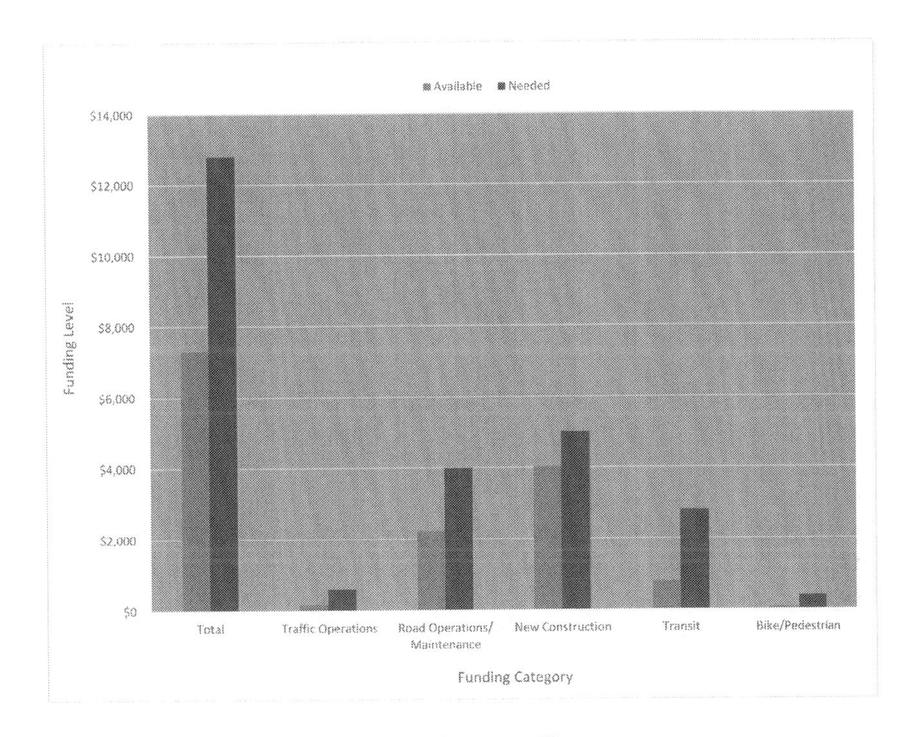

Figure 6.28 Available and Needed Funding in Millions.
(based on Table E-1, Pikes Peak Area Council of Governments, 2008, p. E-1).

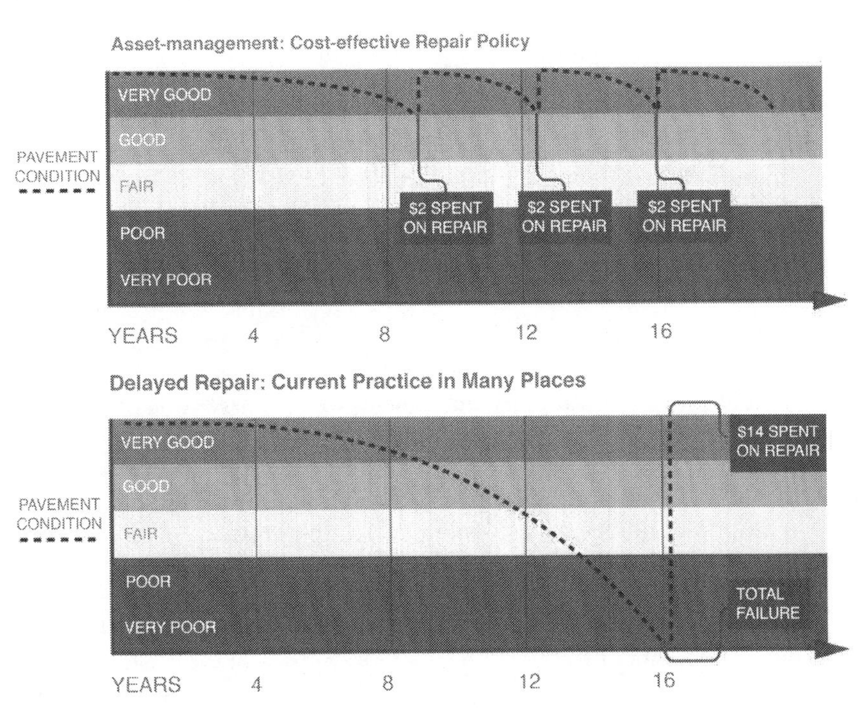

Figure 6.29 Cost Effect of Deferred Maintenance.
(Pikes Peak Area Council of Governments, 2012, p. E-20).

The 2010 Kansas City RTP gives highway maintenance "increased priority." However, a preliminary analysis performed by MPO staff indicates that maintenance spending is still far below the optimum level, and the Greater Kansas City Chamber of Commerce estimates the backlog of deferred maintenance needs at $520 million (Mid-America Regional Council, 2010, p. 4–4). At the same time, the region is funding dozens of new highway capacity projects.

Best practice

The best practice is fix-it-first in word and deed. The reasons are twofold. First, roads in poor condition damage tires and suspension systems, increase fuel consumption, and increase travel time.

The American Association of State Highway and Transportation Officials' (AASHTO) *Rough Roads Ahead* report estimates that poor road conditions in the Greater Philadelphia region cost the average driver $525 per year in additional vehicle expenses (Delaware Valley Regional Planning Council, 2009, p. 6). According to a 2005 report card on Maryland's infrastructure by the American Society of Civil Engineers (ASCE), 45 percent of Maryland's major roads are in poor or mediocre condition. Driving on roads in need of repair costs Maryland motorists $1.4 billion a year in extra vehicle repairs and operating costs (Baltimore Regional Transportation Board, 2007, p. 95).

The second reason is illustrated in the accompanying quote from Los Angeles. Deferred maintenance leads to higher costs in the long run, as roads that could have been patched or resurfaced ultimately need to be rehabilitated or reconstructed.

> Whereas pavement surface damage requires an investment of $64,000 per lane-mile to bring it to a state of good repair, the costs escalate significantly if these investments are not secured in a timely manner. The costs for minor damage repair escalate more than fivefold to $387,000, and the costs for major damage repair escalate to an astronomical $900,000 per lane-mile.
> (Southern California Association of Governments, 2008, p. 13)

Four RTPs represent varying commitments to fix-it-first. At least on paper, the most unequivocal commitment is that of the 2007 St. Louis RTP.

> The Council was progressive nearly 15 years ago when it established preservation as the region's top priority for transportation investment. The region's stakeholders and political leaders recognized how integral a high quality transportation system is to maintaining the social and economic vitality of our region. Since that time, the region has made great strides in not only maintaining, but improving the condition of the existing system.
> (East-West Gateway Council of Governments, 2007, p. 20)

In the 2007 plan, 70 percent of the region's highway budget was programmed for maintenance and operation, whereas only 15 percent went to system expansion (East-West Gateway Council of Governments, 2007, p. 29).

The San Francisco RTP has a somewhat less ambitious policy, but one that is impressive nonetheless in its use of discretionary funds for local street maintenance:

> This plan not only reaffirms the region's long-standing "fix it first" maintenance policy but also expands our commitment to maintaining and operating our existing local roadway and transit systems. The Transportation 2035 Plan directs $7 billion in discretionary funds to maintain local roadways at current pavement conditions, and $6.4 billion to close funding shortfalls for the highest-rated transit assets.
>
> (Metropolitan Transportation Commission, 2009, p. 14)

The $7 billion is enough only to maintain the current state of roadway repair, at which about 22 percent of local roadways are in poor or failed condition. The performance objective chosen for local roadway maintenance is more ambitious than current funding allows – to reduce to 13 percent the share of local roadways in poor or failed condition (Metropolitan Transportation Commission, 2009, p. 28).

The third example of fix-it-first is the Philadelphia RTP.

> Long-range plan policy prioritizes preservation and maintenance needs, followed by operational improvements, then system expansion projects. This approach follows the lead of federal and both state departments of transportation. This "fix-it-first" policy allocates more funding to preserving and maintaining existing roadway and transit networks. The goal is to achieve and maintain a state-of-good repair for existing transportation infrastructure.
>
> (Delaware Valley Regional Planning Council, 2013, p. 99)

Almost 80 percent of highway funds in the 2013 plan are devoted to system preservation. Following a fix-it-first policy, the region has been able to make progress on pavement condition. From 2005 to 2007, the number of lane miles rated as deficient in the region has decreased by 259 miles, or 4.2 percent of the total (Delaware Valley Regional Planning Council, 2009, p. 34).

Finally, there is the 2006 Honolulu RTP, which budgeted $1 billion for "highway system preservation" beyond the $1.38 billion for highway operations and routine maintenance:

> In order to counter some of the neglect of the past, the plan increases spending for system preservation in the early years, then reduces the amount of spending in later years back to traditional levels.
>
> (Oahu Metropolitan Planning Organization, 2006, p. 14)

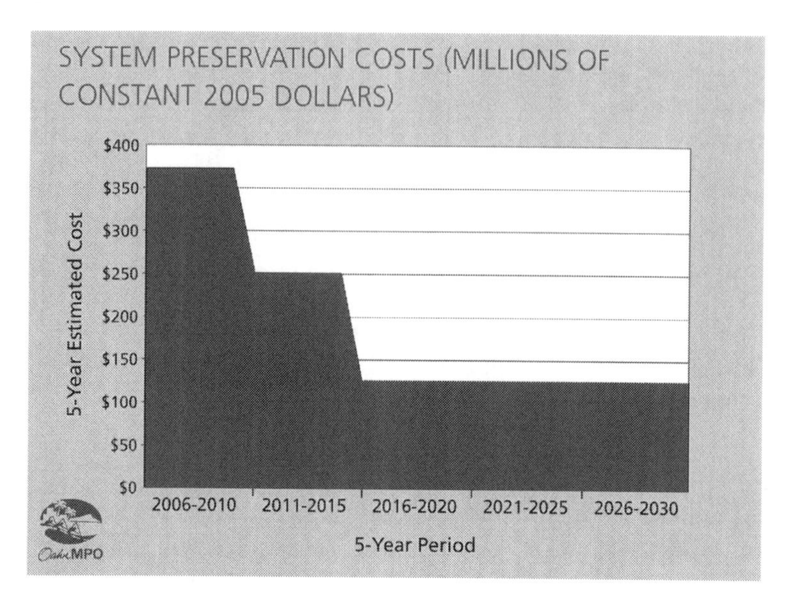

Figure 6.30 Making up for Deferred Maintenance.
(Oahu Metropolitan Planning Organization, 2006, p. 14).

The 2011 RTP continues this policy by allocating over $13 billion, or 56 percent of the overall budget, to support operations, maintenance, preservation, and safety of the existing transportation system (Oahu Metropolitan Planning Organization, 2011, p. 18).

Transportation System Management (TSM) and Intelligent Transportation Systems (ITS)

Background

Transportation system management (TSM) and its cousin, intelligent transportation system (ITS) technology, are broadly defined as measures that make the most efficient use of existing capacity, as opposed to adding capacity. They are also referred to as "operational strategies." "Systems Management helps get the most efficiency out of our existing system, makes travel services more reliable, convenient, and safe, and reduces traffic delays caused by accidents and incidents" (San Diego Association of Governments, 2007, p. 1–8). Relative to highway capacity expansion, the advantages of TSM and ITS are their low cost, short implementation period, and minimal environmental impact.

In its early days, TSM emphasized the reallocation of space within the road right-of-way to better accommodate peak traffic movements. Methods

Table 6.3 General Comparison of Operational Strategies to New Construction

Characteristic	*Construction*	*Operational Strategies*
System Coverage	Location-specific	Network Approach
Cost	Typically High	Comparatively Low
Implementation Timeframe	Long	Short
Responsiveness to Changes in Demand	Low	High

Source: Adapted from Pima Association of Governments (2010, p. 52).

Compared to major construction, operation strategies can be quicker to implement, less costly, more responsive to changes in demand, and provide broader system-wide benefits.

of reallocation included re-striping pavement to create additional (though narrower) travel lanes, removing medians to create turn lanes or additional through lanes, removing on-street parking entirely or during peak hours, converting to one-way streets, and creating reversible median lanes going one direction in the morning, the other in the afternoon. However, each of these measures comes at a cost, in the form of higher crash rates, slower travel in the off-peak direction, less direct routing, and/or less pedestrian-friendly streets. For these reasons, other TSM measures currently gain more attention.

Access management aims to regulate the number of conflict points along a street, as arterials and collectors often become so cluttered with driveways and other access points that they function more like local streets. Research has shown that travel speeds fall, and crash rates rise, as the number of access points increases (Committee on Access Management, 2003). To counter this tendency, comprehensive access management systems establish minimum separations between driveways, traffic signals, and median openings; place restrictions on turning movements into and out of properties; and require turn/acceleration/deceleration lanes where necessary to avoid conflicts with through-traffic. On arterials, four primary approaches consistently achieve traffic flow improvement: access spacing (distancing traffic signals), driveway spacing, dedicated turning lanes, and median treatments such as including two-way left-turn lanes and raised medians (FHWA, 2003). Access management can also benefit cyclists for the same reason it affects overall traffic flow. Without the distraction of constant driveways and cross-traffic, cycling is safer and more comfortable.

ITS includes measures such as electronic detection and prompt removal of crashes and other incidents that block traffic flow. Other ITS systems provide real-time information on traffic conditions (allowing travelers to avoid traffic jams), coordinated traffic signals for optimal traffic flows, and freeway on-ramps that are metered to optimize traffic flows on the mainline. The future promises a more widespread application of these dynamic responses to traffic congestion, combining ITS strategies to create “smart highways.”

Traffic signal systems in many metropolitan areas operate in time-of-day mode: timing plans are activated at different times of day based on normal traffic conditions. When conditions are abnormal, these systems cannot respond. Traffic-responsive systems monitor traffic at critical intersections and select the most appropriate timing plans. Traffic-adaptive systems go even further, continually adjusting and optimizing signal settings as traffic fluctuates, anticipating gridlock before it occurs.

Heavy investment in traffic-responsive and traffic-adaptive signal timing technology over the past decade has greatly improved the state of signal coordination in the United States: 64 of 72 recently surveyed metropolitan areas integrated this technology in at least 20 percent of intersections (RITA, 2007).

Nearly two-thirds of all freeway delay is due to crashes and other incidents (Lindley, 1987). Termed "nonrecurring," this delay seems to recur almost daily as lanes have to be closed and traffic is reduced to a crawl. To respond faster, freeways in many U.S. cities now have incident management systems, complete with traffic detectors, surveillance cameras, electronic advisory signs, and incident management teams. As assessed in the 2009 Urban Mobility Report, incident management systems offer the greatest reduction in delay compared to the most common operational improvements (ramp metering, signal coordination, and access management). Incident management systems offer more than twice the reduction in delay provided by the next-best treatment, access management, and almost three times the reduction provided by each of the other treatments (Schrank & Lomax, 2009). When modeled as if all the treatments were in place on all roads, incident management remains the clear leader in terms of delay reduction.

Another element in corridor management, electronic payment systems use communication technologies to facilitate payments by roadway users, typically for tolls and transit fares. These systems reduce revenue collection and maintenance costs, increase security, and allow for increased throughput. As a result, these systems generally improve facility performance, particularly in systems with smart cards. The Florida Turnpike Authority found a benefit/cost ratio of over 3:1 by increasing numbers of vehicles equipped with electronic toll collection (ETC) technologies, and a 2:1 benefit/cost ratio with just 10 percent of vehicles using the technology in one demonstration.

Traveler information technologies, also known as roadway information systems, include websites, telephone hotlines, television channels, radio frequencies, and signage that distribute information to travelers. These provisions support informed decisions regarding trip departures, routes, and modes of travel. After regional implementation in Cincinnati and Northern Kentucky, 56 percent of surveyed area residents changed their morning routes based on the availability of traffic information. The program also showed a reduction of between 3.6 and 4.7 percent in vehicle emissions, and a 3.2 percent reduction in fatalities attributable to improved response times (GAO, 2005).

A number of challenges currently face the adoption of ITS systems. First, ITS technologies must work within current systems and programs, which often center around outmoded technologies. Through a process known as "mainstreaming," ITS programs seek to build support and acceptance through collaboration with multiple agencies and demonstrating and documenting the benefits of ITS systems. This engagement of multiple cross-jurisdictional agencies proves fundamental to program success, as the programs achieve the highest returns on the regional scale.

Conventional practice

The conventional practice is to label TSM/ITS "cost effective" and a "priority." Yet, when it comes to funding, TSM and ITS often come up short. The two together seldom receive more than 2 percent of total funding.

The Denver RTP states:

> Management and operational strategies (see Chapter 4, Section I) to improve the efficiency and reliability of the roadway system are very important in light of the limited revenues that will be available for expansion of the system.
>
> (Denver Regional Council of Governments, 2011, p. 122)

However, the RTP goes on to state that "anticipated management and operational expenditures cover only about half of the identified need" (Denver Regional Council of Governments, 2011, p. 122). Moreover, the Denver RTP devotes 15 times as much funding to road expansion as to system management and operational strategies (Denver Regional Council of Governments, 2011, p. 107).

The 2008 Southern California Association of Governments' RTP states that:

> In all parts of the region, operational and technological improvements have the potential to maximize system productivity in a more cost-effective way than simply adding capacity.
>
> (Southern California Association of Governments, 2008, p. 104)

For this reason, the RTP made up 20 percent of the original shortfall in TSM funding with diverted funds. This incremental investment of over $2 billion in advanced operational strategies on freeways and arterials was projected to recapture 20 percent of the lost productivity due to congestion (Southern California Association of Governments, 2008, p. 170).

Nonetheless, this left 80 percent of the original shortfall unfunded (Southern California Association of Governments, 2008, p. 201). Moreover, the

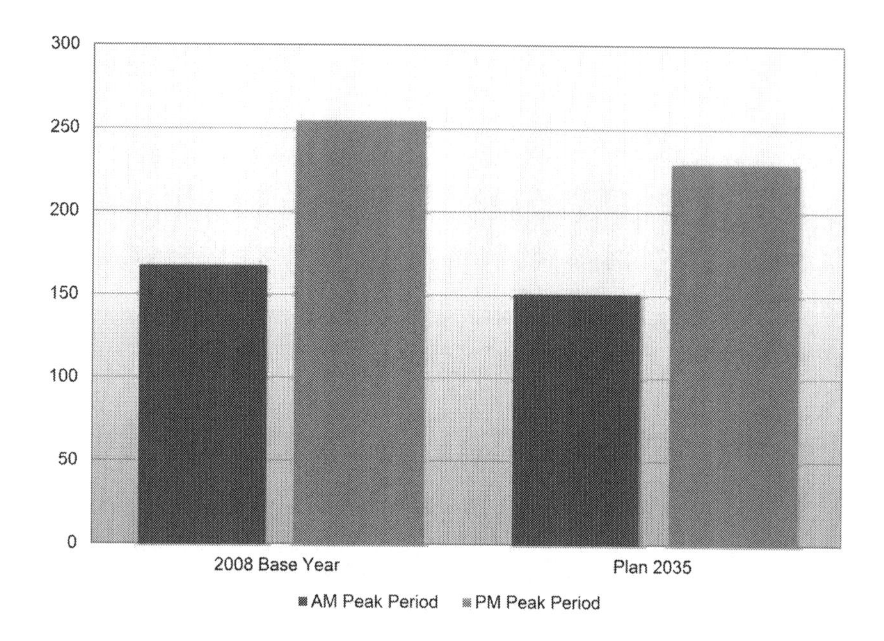

Figure 6.31 Recapturing "Lost Lane-Miles" through an Incremental Investment in TSM.

(Southern California Association of Governments, 2012, p. 176).

RTP earmarked 20 times as much funding for general purpose freeway and arterial lane expansion as for TSM-type activities (Southern California Association of Governments, 2008, p. 161). The RTP's justification:

> As these allocations are programmed and implemented, SCAG hopes that the benefits will become apparent to decision-makers and the public, and that additional funding can be secured to address the remaining shortfall.
> (Southern California Association of Governments, 2008, p. 19)

In the 2012 RTP, the incremental investment to implement advanced operational strategies on freeways and arterials is increased to $6.2 billion. It is still projected to recapture only 20 percent of the lost productivity. One has to wonder why the increment isn't large enough to capture some of the other 80 percent of lost lane-miles.

Best practice

The best practice is to make TSM a priority, and fund it accordingly. The 2008 Atlanta RTP did exactly that. System management, optimization, and

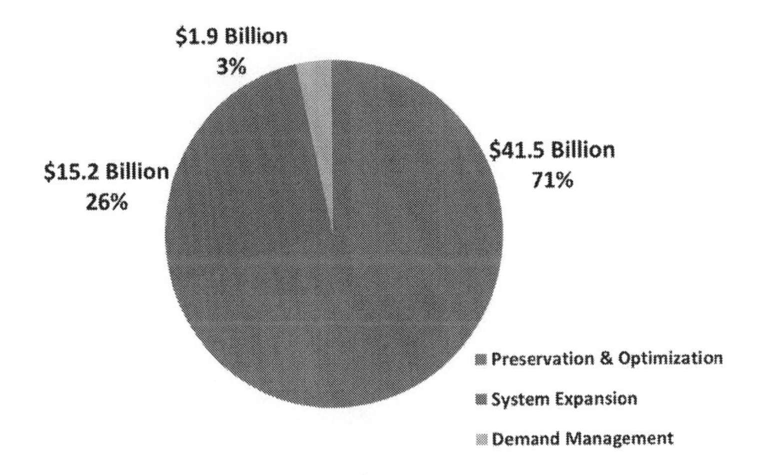

Figure 6.32 Priority Order.
(Atlanta Regional Commission, 2014, p. 4–3).

operations (which includes highway and transit O&M) were the top priority of the MPO. The category of "non-capacity roadway optimization" received more than 5 percent of total funds, more than in any other RTP. Notably, some of this focus was lost in the 2014 Atlanta RTP.

Whereas it is beyond the scope of this review to sort through the list of TSM measures and identify the most cost-effective ones, we will make certain observations based on data in the RTPs. Several of the RTPs cite the Federal Highway Administration finding that most congestion on freeways is "nonrecurring," that is, due to incidents that do not occur on a daily basis.

> Only 40 percent of congestion is due to recurring delay (routine commute traffic). The Plan 2040 RTP (March 2014 Update) allocates $3.5 billion of investment to target the other 60%. TSM&O investments "take back" the capacity lost of non-recurring events such as incidents, bad weather, and inadequate signal timing . . . . These types of strategies are notorious for being more cost effective than capacity-adding investments because they are much cheaper and quicker to implement or maintain . . . . Many of these programs result in benefit-to-cost ratios as high as 16 to 1.
>
> (Atlanta Regional Commission, 2014, p. 4–7)

To deal with nonrecurring congestion, MPOs fund incident management systems that pre-position incident response personnel and equipment and use advanced technology to detect and verify traffic incidents. These measures improve response time and allow for quicker removal of incidents and restoration of traffic flow.

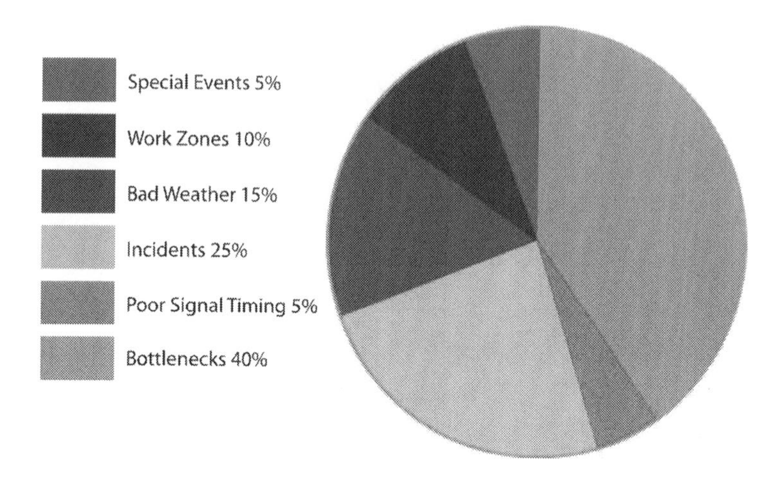

Figure 6.33 Nationwide Causes of Delay.
(Baltimore Regional Transportation Board, 2011, p. 4–29).

As for recurring freeway congestion, a measure that would seem to be particularly cost-effective is the installation of freeway ramp meters to ensure that merging traffic does not exceed the merge area or weave area's capacity to absorb that traffic at a point. Ramp meters spread out the entering vehicles. Ramp meters are also used to control overall flow to assure that downstream traffic flow remains stable. Figure 6.34 from the San Francisco RTP shows the tremendous reduction in travel time with the introduction of ramp meters.

As for recurring congestion on the arterial system, nearly all delay on surface streets occurs at intersections, so any measure that reduces intersection delay will do more for travel speeds than any action at midblock. At moderate travel volumes, roundabouts produce less intersection delay than do traffic signals, and several RTPs call for the use of roundabouts in such cases (Capital Area Metropolitan Planning Organization of Austin, 2010, p. 11; Ohio-Kentucky-Indiana Council of Governments, 2008, p. 8–6; Capital Area Metropolitan Planning Organization and Durham-Chapel Hill-Carrboro Metropolitan Planning Organization, 2013, p. 71; Tri-County Regional Planning Commission, 2010, p. 8–9). The Utah Department of Transportation (UDOT) has been particularly aggressive in pursuing other innovative intersection designs such as continuous flow intersections and ThrU Turn intersections (Wasatch Front Regional Council, 2011).

At higher traffic volumes or with imbalanced flows, signals are likely to outperform roundabouts, provided that they are optimally timed. Many RTPs earmark funds for signal coordination and synchronization. A few

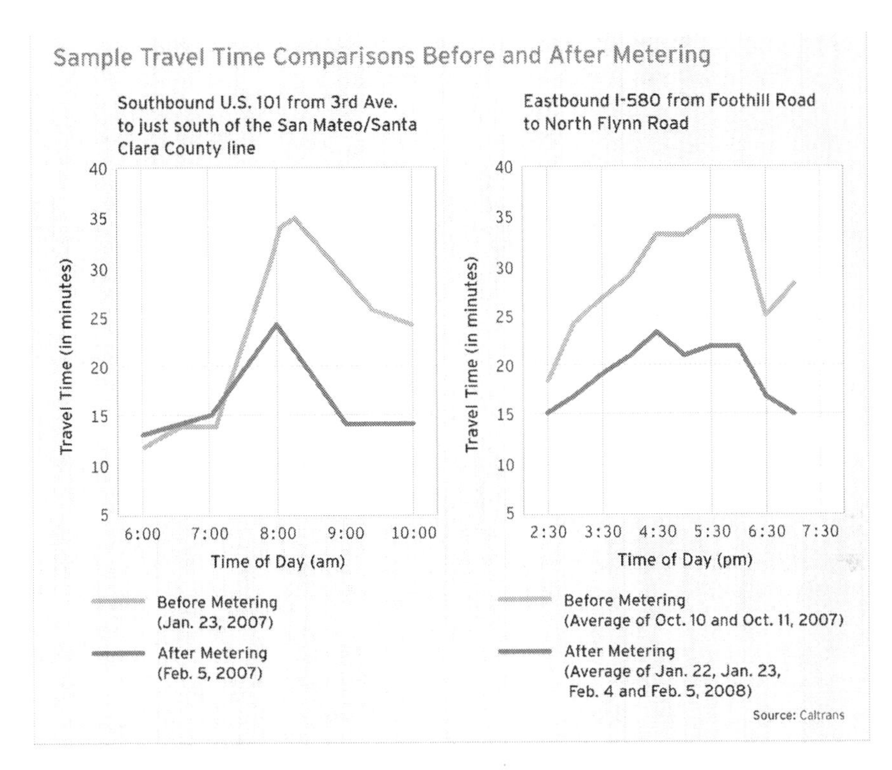

Figure 6.34 Sample Travel Time Comparison before and after Ramp Metering.
(Metropolitan Transportation Commission, 2009, p. 55).

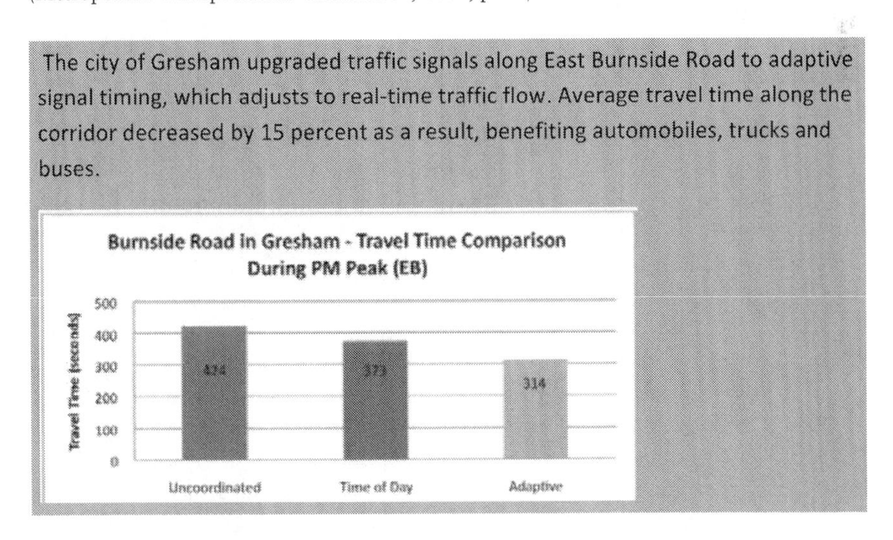

Figure 6.35 Comparison of Travel Times under Uncoordinated, Time of Day, and Adaptive Traffic Control.
(Portland Metro Regional Government, 2010, p. 2–79).

go beyond the standard technology in the United States to endorse traffic-responsive and traffic-adaptive signal control.

Traffic-adaptive systems have been used in Britain and Australia for decades, and have begun to be deployed in the United States, with impressive results. The Portland RTP reports that deployment of adaptive traffic control in the city of Gresham resulted in a 15 percent reduction in travel time in the test corridor, "benefiting automobiles, trucks and buses" (Portland Metro Regional Government, 2010, p. 2–79).

Travel Demand Management (TDM)

Background

In discussing the range of transportation investment options, the majority of strategies address the supply side of travel behavior. Here we consider the demand side, addressing programs known collectively as Travel Demand Management (TDM). There is evidence that congestion results primarily from bottlenecks in traffic caused by an overload of drivers, rather than by other factors. TDM seeks to address this overload, rather than pursuing marginal improvements from roadway systems investments. Modern TDM now relies heavily on employer incentives, financial incentives, and strategies tailored to different markets. Some TDM strategies also support dense development, whereas others improve access to sites, and still others address parking supply and usage.

Travel demand management strategies reduce the demand for peak-period single-occupant vehicle travel by (1) promoting alternatives to driving alone, (2) shifting trips out of peak travel periods, or (3) eliminating the need for certain trips. The Chicago RTP lists the following strategies:

- Ridesharing Programs. Ridesharing can reduce congestion by reducing the number of vehicle trips, in turn leading to reductions in VMT.
- Car Sharing Programs. Car sharing reduces VMT by reducing vehicle ownership; cars are available when needed, but discretionary trips may be more likely made by transit or non-motorized modes.
- Alternative Work Arrangements. Alternative work arrangements reduce VMT by providing work sites closer to homes, or by spreading traffic to non-peak periods.
- Transit and Rideshare Incentives. Economic incentives for transit and ridesharing can reduce the costs of these modes, encourage their use, and thus reduce VMT.
- Parking Management. Parking management manages the cost of parking, reduces its availability, provides information regarding availability, so as to reduce travel demand and reduce excess VMT searching for parking spaces.

- Guaranteed Ride Home Programs. Guaranteed ride home programs reduce VMT through increased transit use by assuring transit users a way home should they need to travel when transit is not available.

(Chicago Metropolitan Agency for Planning, 2008, p. 72; 2010, p. 358)

Employers are the key to TDM success. They have the ability to modify work hours and establish telecommuting programs. They can provide financial and other rewards to their employees who rideshare. They can hire transportation coordinators to personalize ridematching, train telecommuters, and run vanpool programs. In a review by Kuzmyak et al. (2010), employers providing transportation services such as carpool and vanpool programs saw a vehicle trip reduction (VTR), based on average daily vehicle trips, of 24.2 percent to 26.5 percent.

Common support actions provided by institutions or employers include hiring employee transportation coordinators, providing on-site transit information and pass sales, providing preferential parking for carpools, and implementing guaranteed ride home programs. Guaranteed ride home (GRH) programs, for example, provide a safety net for commuters who forego their personal vehicle and may need to work late or return home in an emergency. Such programs commonly reimburse employees for taxi fares, or allow use of company vehicles. In one study, the Metropolitan Washington Council of Governments (MWCOG) found that 31 percent of commuters felt that GRH programs were important in their decision to use an alternative to driving alone (MWCOG, 2002). For bicycle commuters, support actions include providing facilities such as bicycle storage, lockers, showers and changing facilities.

In a 2004 National Center for Transit Research study, the most frequently observed support actions consisted of rideshare matching, guaranteed ride home programs, and bicycling amenities. These programs serve as enhancements to financial incentives and physical services, and prove most effective when offered in conjunction with these strategies (CUTR, 2004).

Employers can be mobilized through development agreements/orders (DAs/DOs), transportation management associations/organizations (TMAs/TMOs), or trip reduction ordinances (TROs).[1] DAs/DOs are agreements reached at the time of development approval requiring specific actions on the part of developers or their successors. TMAs/TMOs are organizations set up by regions or groups of employers to coordinate their TDM programs. They provide ridematching services, sell discount transit passes, run shuttles, and perform other mobility-related functions for their members. TROs are ordinances that require employers to reduce peak-hour vehicle trips to their sites. Whereas the choice of TDM measures is usually left to employers, trip reduction targets are set for them. If they fall short, employers must offer more incentives to their employees until the targets are met. By the best estimates now available, employer trip reduction measures appear very competitive with other means of reducing VMT and vehicle emissions (Schreffler, 1996).

Parking charges and/or parking limitations represent the most effective ridesharing incentives. They have been called the "biggest lever available to employers and institutions" for reducing travel demand (Kuzmyak et al., 2010). Solo commuting can be reduced by 20 percent or more by instituting market-rate parking charges. Employees have great incentive to share a ride or use transit when several dollars in parking charges are added to their daily out-of-pocket costs. The review of TDM strategies by Kuzmyak et al. (2010) shows a 22.8 percent to 24.1 percent reduction in vehicle trips for employers with parking fees or parking restrictions, compared to 12.2 percent for employers without these provisions. Furthermore, analysis of combined effects shows that parking supply and pricing have the greatest impact on the effectiveness of all other TDM program combinations. The incentive for employees becomes even greater when ridesharers receive free or discounted parking while solo commuters pay the full price.

To overcome employee resistance, employers may introduce travel allowances at the same time as parking charges; employees who solo commute will break even while those who rideshare can pocket the difference (Ehrlich, 1996). Alternatively, employers may offer employees the option of a free parking spot or an equivalent mass transit or ridesharing subsidy. Called parking "cash-out" subsidies, a limited version of this policy now applies to all employers with more than 50 employees who provide subsidized or leased parking spaces for their employees in California nonattainment air basins (California Air Resources Board, 1991). The effects of these arrangements are well-established. In one evaluation, the share of solo commuters dropped from 76 to 53 percent when cash payments were offered in lieu of parking subsidies (Shoup, 1997). In another example, employers maintained free parking without additional charges, but offered travel allowances for those choosing not to drive alone. This program, a form of cash-out, reduced the number of commuters driving alone from 69 percent to 55 percent (Shoup, 1997). More recent evidence, however, shows that this law may not be as effective as initially thought (Weikel, 2015). In addition, this law only applies to approximately 3 percent of the 11 million free parking spaces statewide (Weikel, 2015).

Beyond cash-out programs, employers may offer options such as transit subsidies and vanpool subsidies. According to Kuzmyak et al. (2010), benefits such as prepaid transit passes and vanpool financing show a 16.9 percent to 20.5 percent rate of vehicle trip reduction. Additionally, programs offering subsidies in combination with transportation services exhibit greater reductions in vehicle trips of up to 26.7 percent. In Ann Arbor, Michigan, a discounted/free transit pass program by itself reduced vehicle trips by 3.5 percent whereas bus passenger trips increased by 9.2 percent (Association for Commuter Transportation, Urbantrans Consultants, PB Parsons Brinckerhoff, & ESTC, 2004). At the University of California (Los Angeles), the provision of an unlimited pass program increased commuting to campus on transit by 56 percent while decreasing driving alone by 20 percent (Georggi, Winters, Rai, & Zhou, 2007).

Beyond financial incentives, TDM programs benefit by managing the supply of parking, either through areawide approaches or site-specific management. Areawide parking supply management strategies regularly address minimum and maximum parking requirements and parking ratio policies. Portland, Oregon, has used maximum parking ratios, on-street parking management, and a cap on downtown parking since 1975, with adjustments in 1995 to address VMT growth and manage areas surrounding the CBD. All of these strategies have been credited with contributing to high transit mode shares in the downtown (43 percent), as well as high carpool rates (Kuzmyak et al., 2010). An additional supply management strategy provides preferential parking for exclusive use by HOVs, with reserved spaces in preferred locations such as covered spaces, on-site locations, or spaces near building entrances (Kuzmyak et al., 2010). This strategy proves effective primarily as one component within a suite of programs, rather than as a stand-alone incentive (Pratt, Pedersen, & Mather, 1977; K.T. Analytics, 1995).

A review of 82 programs by Kuzmyak et al. (2010) showed that sites offering telecommuting may reduce vehicle trips between 16.6 percent and 28.2 percent, and a compressed workweek schedule may reduce vehicle trips by 19.7 percent. Regarding the marginal effects of these approaches when combined with other strategies, the review shows that only the flexible hours generate significant additional VTR benefits. Comparing programs with and without flexible hours, the review showed a 7 percent greater VTR for programs that included flextime schedules. On this basis, altered work schedules and telecommuting may benefit specific employers, while playing a supporting role on a broader scale.

Regions may move from a focus on employer TDM programs to regional policy. An areawide tax or surcharge, for example, may apply to both on-street and off-street parking. In 1970, San Francisco applied a 25 percent parking tax to all off-street parking. Traveler response showed a parking price/parking demand elasticity of –0.3. Thus, for every unit of increase in parking price, parking demand decreased by 0.3. However, subsequent analysis showed that behavior changes appeared in more subtle forms than simple parking demand. For example, shoppers shortened the duration of their parking times, whereas many daily commuters stopped parking in expensive facilities and parking garages (Kulash & Urban Institute, 1974).

TDM focuses on commuters because their trips are concentrated in peak hours and repeated daily. But with work trips representing less than a fifth of all daily trips, and only about half of all trips made during afternoon peak hours, TDM impacts will be limited as long as TDM applications are limited. Options for managing nonwork trips (in order of increasing potential) include computerized carpooling programs; neighborhood TDM programs;

restrictions on truck delivery/pickup times; advanced traveler information systems diverting travelers to less congested routes, modes, and times; and if the political resistance can ever be overcome, congestion pricing programs that charge motorists for use of roads in congested times or places. Areawide parking pricing, discussed above in terms of areawide and regional applications, also influences nonwork trips, shifting the demand for parking in central areas and during peak travel periods.

Conventional practice

Without exception, the emphasis in RTPs is on increasing the supply of transportation capacity, not reducing the demand for transportation services. As the Raleigh and Durham RTPs note:

> Each year, hundreds of millions of dollars are spent in the region on the supply side of mobility: building and maintaining roads, buying and operating buses, building sidewalks and bicycle facilities. (Yet) some of the most cost-effective mobility investments we can make are on the demand side: encouraging commuters to use our transportation facilities as efficiently as possible by carpooling, vanpooling, taking transit, telecommuting or walking or bicycling.
>
> Capital Area Metropolitan Planning Organization and Durham-Chapel Hill-Carrboro Metropolitan Planning Organization, 2013, p. 59)

In the early 1990s, local trip reduction ordinances, requirements of the California South Coast Air Quality Management District's Regulation XV (Rule 1501), and the Clean Air Act Amendments of 1990 (CAAA) made TDM programs mandatory. Employer opposition caused several states to suspend their programs. Federal legislation in late 1995 allowed states to implement programs on a voluntary basis. California, the early laboratory for such programs, repealed Rule 1501 in December 1995.

Since then, TDM efforts have been fragmented and ad hoc in most metropolitan regions. TDM programs are a rounding error in the typical RTP budget. Of those relatively few plans that break out expenditures for TDM, they almost always represent less than 1 percent of regional transportation expenditures. Moreover, as with many other stated priorities, TDM programs are only partially funded in many RTPs. The Denver RTP is typical in stating the importance of TDM services:

> With limited funding available for expansion of the roadway system, TDM services (see Chapter 4, Section J) will be critical to reducing motor vehicle travel demand and offering mobility options.
>
> (Denver Regional Council of Governments, 2011, p. 122)

Nonetheless, just a little more than half of the envisioned cost of providing TDM services is funded in Denver's fiscally constrained RTP (Denver Regional Council of Governments, 2011, p. 119).

Best practice

A first best practice is to set ambitious but realistic targets for employee trip reduction. The Triangle Region TDM Plan seeks to reduce the growth in the amount of commuter travel by 25 percent (Capital Area Metropolitan Planning Organization and Durham-Chapel Hill-Carrboro Metropolitan Planning Organization, 2013, p. 59).

The second best practice is to fully fund TDM programs within RTPs. "One of the most cost-effective ways to reduce congestion and the demand for new and wider roadways is to promote and support travel demand management" (Pima Association of Governments, 2014, p. 56).

A third best practice is to coordinate regional TDM activities. In the New York region, the New York Metropolitan Area Regional Commuter Choice Program delivers integrated and coordinated alternatives to driving alone to work. The program, sponsored by New York State DOT with the participation of the New York MPO's member agencies, supports commuter assistance services, coordination of online rideshare matching services, and coordination of human service mobility management programs (New York Metropolitan Transportation Council, 2013, p. 4–22).

A fourth best practice is to focus TDM efforts within activity centers, the same centers that are promoted in many RTPs. Washington State amended its Commute Trip Reduction (CTR) Program in 2006 to focus on urban growth areas and centers where TDM is most needed and where it will have the greatest impact on travel demand. In Denver, CO, TDM promotion and facilitation efforts are offered throughout the region but are concentrated in:

- Downtowns of major cities and high employment concentration areas;
- Along highway corridors with bus/HOV lanes; and
- Adjacent to rapid transit lines/stations and high transit service locations in conjunction with major highway construction projects.

(Denver Regional Council of Governments, 2011, p. 96)

A fifth best practice is to make TDM mandatory for large employers (something that requires state or local legislative action). The Oregon Employee Commute Options Rule mandates the Regional Travel Options Program in the Portland region. Oregon was one of only two states (Washington was the other) where the percentage of people driving alone to work decreased between 1993 and 2000 (Portland Metro Regional Government).

Transit expansion

Background

For most of the last century, transportation systems were funded and developed to accommodate the automobile, leaving public transportation in a downward spiral of disinvestment and decreasing ridership. With the current trends of an aging population, increasing congestion, and environmental degradation, the focus is shifting from automobiles to multimodal systems. Transit and cities are now experiencing a renaissance, with transit use increasing faster than automobile travel.

Transit improvements provide benefits that road improvements often cannot, such as cleaner air, greater mobility for our aging society, and environmental sustainability (APTA 2009, p. 3). The analysis in Chapter 5 suggests that each 10 percent increase in transit frequency and in route density result in a 0.6 percent reduction in regional VMT per capita. In contrast, every 10 percent increase in highway lane miles per 1,000 population results in a 1.6 percent increase in VMT per capita. The Texas Transportation Institute (Schrank et al., 2012) found that existing transit generates substantial benefits in terms of reduced hours of traffic delay, reduced fuel consumed, and reduced dollars spent; however, the majority of these benefits accrue to the top 15 largest urbanized areas.

There is a vast literature on transit ridership determinants, led by Brian Taylor of UCLA and Jeffrey Brown of Florida State University. We can only touch on a fraction of it here.

Bus vs. Rail: Historically, bus versus rail debates have been narrowly focused on opinion, lacking adequate supporting materials and evidence (Zhang, 2009). The virtues of bus are generally argued to be cost and flexibility, whereas rail provides more comfortable and reliable service, and offers great potential for redevelopment.

Rail is generally most appropriate in very busy urban corridors with heavy ridership demand. The initial capital costs are high and it can take years to build ridership and affect land use patterns. However, as ridership increases and costs are averaged over the long term, unit costs of rail decrease.

Local and regional rail transit networks, including both light rail and commuter rail, offer a number of benefits both as a supplement to bus service and, in some cases, as alternatives to bus service. Advantages include consistent ridership for both new and existing rail infrastructure, as well as political support for rail as an alternative to highway expansion. Winston and Langer (2004) found that both motorist and truck congestion costs decline in a city as rail transit mileage expands, but congestion costs increase as bus transit mileage expands. This appears to occur because buses attract fewer travelers from driving, contribute to traffic congestion themselves, and have less positive impact on land use accessibility.

Henry and Litman (2006) found that between 1996 and 2003, cities focusing on rail system expansion rather than bus system expansion achieved greater results in transit ridership, passenger-mileage, and operating cost efficiency. Likewise, Bento, Cropper, Mobarak, and Vinha (2003) found that a 10 percent increase in rail supply creates a 40-mile reduction in annual VMT per capita, compared to a 1-mile reduction for a 10 percent increase in bus service. Ewing et al. (2014) found that an extension of light rail to the University of Utah in Salt Lake City, replacing bus service, took 10,000 cars per day off roads in the LRT corridor. Brown et al. (2015) and Ramos et al. (2015) found that streetcar service in one city, Portland, outperformed the average local bus in terms of service productivity and cost effectiveness. While generally viewed as an economic development tool, the authors explain how streetcar service can also excel as a transportation mode.

Bus service is more appropriate at lower densities. Buses have the advantage of customization to meet the changing individual land use and travel patterns of an area. Routing and coverage changes can enhance transit system efficiency and effectiveness, making it a competitive transportation mode. Pratt et al. (2004) note that improved bus routing may serve to reduce time spent on the bus, shorten the walk necessary to reach bus service, or reduce the number of transfers (and transfer time) necessary to make a trip.

Bus improvements can also be a cheaper alternative. The total capital and operating cost of increasing transit ridership with bus routing and coverage enhancements is low when compared to rail solutions. Bus improvements like typical peak period bus services, route restructuring, real-time bus tracker and information systems, feeder services, and reverse commuter

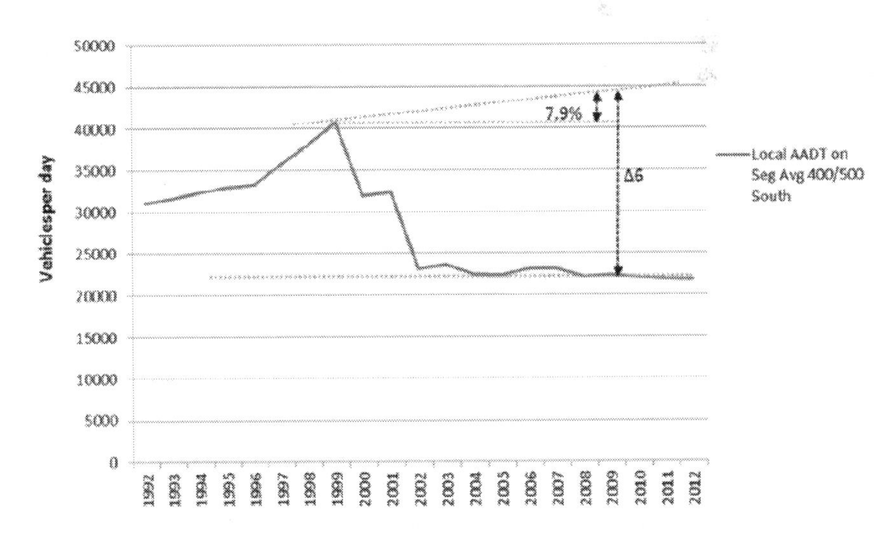

Figure 6.36 Annual Average Daily Traffic on 400/500 S, Estimated Local Traffic on 400/500 Based on Trip Generation Between 1999 and 2009 (LRT extensions in 2001 and 2003).

(Ewing et al., 2014, p. 108)

services can make a difference in service and increase ridership at minimal cost (Pratt et al., 2004, p. 10–54; Tang & Thakuriah, 2012).

As defined by the Federal Transit Authority, Bus Rapid Transit, or BRT, refers to a "rapid mode of transportation that can provide the quality of rail transit and the flexibility of buses" (Danaher, Levinson, & Zimmerman, 2007). More specifically, BRT systems integrate features to improve the speed, reliability, and identity of rubber-tired bus transit. Features include bus-only running ways separated from other traffic, enhanced stations, off-board fare collection, real-time passenger information, and frequent service. BRT systems have been estimated to increase base ridership between 5.8 percent to 6.6 percent based primarily upon improvements in travel times, and potentially up to 25 percent more when implemented as part of a "complete" network incorporating all key features associated with BRT (Danaher et al., 2007). Zhang (2009) suggests that BRT has the capacity to outperform LRT at moderate population and job densities, with lower capital costs and equivalent service levels.

While socioeconomic factors are important influences on transit performance, they are not determinative. Planning and operational decisions also emerge as important influences on LRT and metropolitan transit performance generally (Jaroszynski & Brown, 2014; Taylor & Fink, 2013):

Fares: Pricing can make transit a competitive mode, especially if the trip maker can reduce trip costs by using transit. In TCRP Report 95, Pratt et al. (2004) concluded that if the overall cost of the trip, including transit fare and cost to get to the facility, is less than the cost of the trip using the automobile only, demand tends to be higher. Litman (2007) reports that the overall average elasticity of transit ridership with respect to fares is –0.4, meaning that each 1 percent fare increase will reduce ridership by 0.4 percent and, conversely, each 1 percent fare reduction will increase ridership by 0.4 percent. Similarly, Kain and Liu's (1996) econometric analysis of 184 systems over a 30-year period found that the mean fare elasticities for ridership changes during the 1970 to 1980 and 1980 to 1990 periods ranged from –0.34 to –0.44. The models are consistent with both observation and intuition: high fares drive away passengers and lower fares attract passengers. Fare programs targeted at specific populations can be very effective in attracting riders (Taylor & Fink, 2013).

Transit Scheduling and Frequency: Taylor and Fink (2013) found that improvements in service are more important than reduction in fares. Evans et al. (2004) found that with each 1 percent increase in transit service frequency, vehicle mileage, or operating hours, ridership increases by 0.5 percent. Response to frequency changes tends to be greatest when the prior frequency was less than three buses per hour (Pratt & Bevis, 1971), when a route serves middle- and upper-income areas (Holland, 1974), and when the travel market comprises trips short enough that walking is an option (Evans et al., 2004, p. 9–11). Taylor et al. (2009) found, after controlling for the fact that public transit use is strongly correlated with urbanized area size, that about 26 percent of the observed variance in per capita transit patronage across U.S. urbanized areas is explained in the models by service frequency (p. 62).

Introducing evening, weekend, and off-peak hours can attract riders with nontraditional work hours and make it easier to use transit for discretionary trips. Currie and Loader (2009) found that extending service hours led to higher ridership numbers. Evans et al. (2004) gives a classic example from Reston, Virginia, when a bus was added in 1970 to pick up late passengers in downtown Washington. Ridership on the bus varied between 15 and 20 passengers per trip, but more than 80 new riders were attracted to the system with this improvement that assured them they would not be stranded.

Ride and Walk Times: A study of transit ridership in Atlanta found that transit-dependent bus riders are highly sensitive to travel times, and that choice rail users primarily value fast service and short walk times. The study concluded that an agency could increase ridership by both groups using a core network of higher speed lines that provide access to decentralized employment centers (Brown et al., 2014).

Wait and Transfer Times: An examination of over 50 work purpose travel demand models from throughout the United States found each minute of transit wait time to be 2.12 times as deleterious as a minute of in-vehicle travel time (Evans et al., 2004, p. 9–22). A reduction in Chicago rail rapid transit wait times of 20 percent resulted in a 1.8 percent ridership gain (Evans et al., 2004, p. 9–27).

Conditions while waiting (lighting, cleanliness, information, security) become more important the longer the wait (Yoh et al., 2011). Adverse weather can be a significant predictor of transit ridership (Tang & Thukuriah, 2012), implying that shelters in particular are important.

The cost of transfer penalties can be reduced by increasing frequency and reliability and by improving the physical quality of stops/stations (Iseki and Taylor 2009). Concise and easy to understand information has been shown to reduce actual and perceived transfer costs, especially for novice users (Iseki & Taylor, 2009).

When transit service is infrequent, transfer times can be kept to a minimum, and service reliability increased to its maximum, with timed transfers (Vuchic, Clark, & Molinero, 1981). Several bus routes converge at a transfer center at about the same time, much as connecting airline flights converge at a hub airport. In several cities, conversion to timed transfers has reversed downward trends in transit ridership (Stanley, Gleason, & Kyte, 1982; Schneider, Deffebach, & Cushman, 1984; Bakker et al., 1988). Portland implemented a timed-transfer system in its Westside community in 1979 and saw a 40 percent increase in overall ridership. In certain areas, local trips increased by 138 percent, and nonwork trips increased by 68 percent (Evans et al., 2004, p. 9–18).

Transit Reliability: Evans et al. (2004) cite attitudinal studies of commuters in Baltimore and Philadelphia that found "arrival at intended time" to be perceived as the second most important travel attribute for work trips. Virginia Railway Express (VRE) encountered reliability problems after a freight train derailment in 1996 that caused chronic delays for weeks, along with individual train cancellations (Evans et al., 2004, p. 9–20). Ridership fell by 32 percent in the following months, and although VRE was originally projecting ridership growth for the year, it experienced an additional 16 percent loss (Finn, 1997). Yoh et al. (2011) found that improved schedule adherence is an effective way of reducing transfer burdens and increasing ridership.

Park-and-Ride: Well-placed park-and-ride facilities make transit an easier decision for many commuters (Farhan & Murray, 2008). The fact that they use their auto for one leg of the journey shows that they have an auto and could take it all the way to their destination. In Connecticut, the commuter rail line experienced an increase in ridership from 0.74 to 0.77 riders for every new parking space (Turnbull et al., 2004, pp. 3–12).

Transit Information and Promotion: Providing information and promotional materials can help educate citizens about their travel options and the impacts of their choices. Mass marketing, incentives, and information services are common tactics among transit agencies. The increasing sophistication and diffusion of information technologies improve user experience by providing real-time information on transit arrivals, thus reducing anxiety and perceived wait-time (Iseki & Taylor, 2009). Binary logit and ordered probit models in Santa Clarita and Sacramento found 59 percent of respondents were likely to use transit at least once per week given the availability of Intelligent Transportation System (ITS) delivered transit information, and about half of the nontransit users who might consider transit would be more likely to use it if certain information items were available (Abdel-Aty & Jovanis, 1995, cited in Taylor & Fink, 2003). In addition, perceived wait time for users without real-time information is higher, and real-time information users report a wait time of 7.5 minutes compared to 9.9 minutes for those riders using traditional schedule information (Watkins et al., 2011).

Bike Accommodations: Studies of bike accommodations show similar results as it is mutually beneficial to integrate bicycle infrastructure and planning with public transportation (Pucher & Buehler, 2009). Bike carriers on transit vehicles allow bicycles to be used not only for access to transit but for egress at the other end of trips; easy access and egress is something even the automobile (in a park-and-ride mode) cannot match. Parking, bicycle stations with access to repair tools, paths that lead to public transportation stations, and the ability to bring bicycles on board public transportation vehicles (i.e., light rail car) have all contributed to the integration of bike use and transit (Pucher & Buehler, 2009).

Service Integration: Metropolitan areas that have integrated rail transit into a decentralized network of bus routes have been found to enjoy higher riding habit, higher service productivity, and better cost-effectiveness than metropolitan areas with other network structures or modal combinations (Brown & Thompson, 2008, 2009).

Conventional practice

The conventional practice is to fund highways heavily, and transit lightly, and then lament the fact that travelers seem to prefer the automobile. Consider the following passages from the 2005 Kansas City RTP.

> Kansas City's system of roadways is among the most extensive in the nation. Recently, new statistics made available from the Federal Highway Administration confirm that Kansas City continues to possess the most freeway miles per person of all urbanized areas with populations greater than 500,000.
>
> (p. 2–12)

> Travelers in Greater Kansas City take advantage of the short travel times and low levels of congestion by traveling more miles per person each day than residents of other metropolitan areas.
>
> (p. 2–21)

> It should be noted that the sharp drop in the KCATA's [transit] ridership during the early 1980s is in part due to the significant reduction in service implemented at that time . . . ridership has decreased slightly since 2001, mostly due to a reduction in scheduled service miles resulting from decreased funding support.
>
> (p. 10–9)

> much of the region's [transit] services, especially in lower-density areas, is provided in the peak period only, with many areas having no off-peak services.
>
> (p. 5–4)

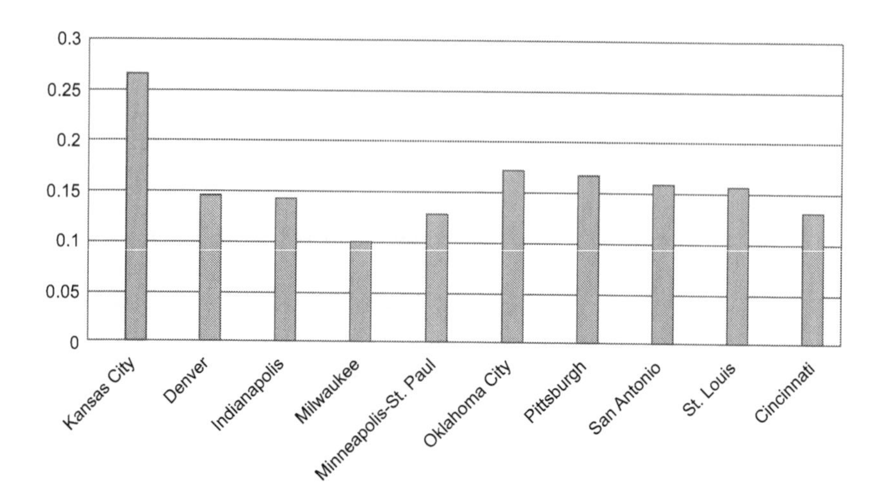

Figure 6.37 Freeway Capacity Comparison: Freeway Miles per 1,000 Persons. (Mid-America Regional Council, 2005, p. 2–12).

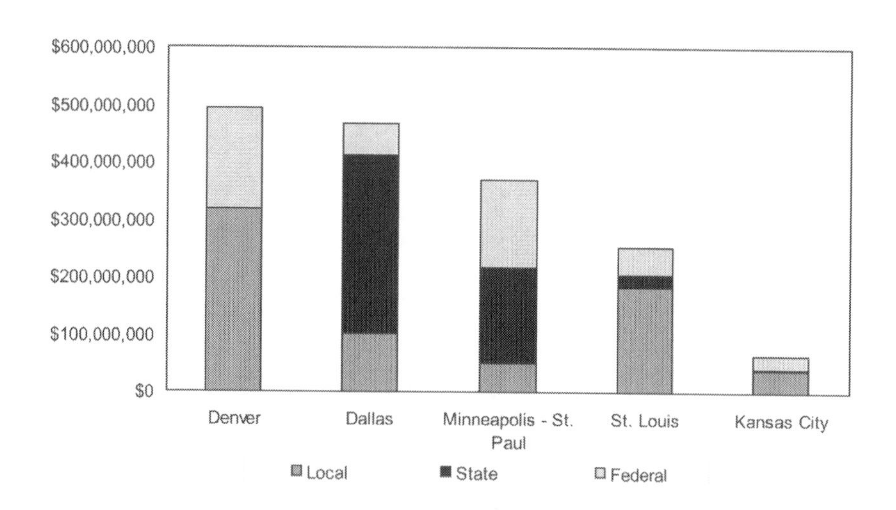

Figure 6.38 Transit Funding Sources Comparison. (Mid-America Regional Council, 2005, p. 5–26).

It is a self-fulfilling prophecy that transit cannot compete with the automobile if transit service is cut and freeways are expanded. Only 1.3 percent of commute trips in Kansas City were by transit according to the 2005 RTP, one of the lowest shares for a large urbanized area in the United States (p. 2–15).

The focus on disproportionate highway funding is changing (Mid-America Regional Council, 2015, p. 4.1). In the most recent RTP, the goals around public transportation include (1) support and sustain existing services, (2) create a regional mobility management system, (3) expand regional public transit and management services, (4) integrate public transit and enhanced mobility services into the built and natural environment, and (5) provide accurate and up-to-date information about existing and planned services (Mid-America Regional Council, 2015, pp. 5.26–5.28).

Best practice

The best practice gives transit investments as much priority as highway investments, because transit investments can ease traffic congestion without inducing additional VMT and the social costs that accompany it (greenhouse gas emissions, for example). An average Boston resident drives 21–23 percent less than the average Massachusetts resident and 30–35 percent less than the average American due in part to its public transportation system (Boston Region Metropolitan Planning Organization, 2011, p. 5–29). The Texas Transportation Institute has performed similar calculations, with similar results, for other large regions.

Because it represents a win-win for congestion and VMT, transit gets more funding than its mode share might suggest. In San Francisco:

> Buses, trains, ferries, light-rail vehicles, cable cars and streetcars not only provide mobility for people without cars – including those who are low-income, elderly, disabled or too young to drive – they also provide a viable alternative to driving for hundreds of thousands of area residents who do own cars. By reducing the number of vehicles on the roads, public transit helps to fight congestion and curb greenhouse gas emissions.
>
> (Metropolitan Transportation Commission, 2013, p. 43)

In Boston, the latest RTP indicates that $190 million from the MPO Discretionary Capital Program will be “flexed” from highway funding to transit (Boston Region Metropolitan Planning Organization, 2015, p. ES-9), and public outreach shows a preference for the MPO to allocate 25 percent of its budget, on average, to the Flex to Transit program (Boston Region Metropolitan Planning Organization, 2015, p. 2–18).

Even in Los Angeles, with its autocentric culture, more is now being spent on transit than on highways.

> Despite a common perception of an auto-oriented culture, the region’s transit system includes an extensive network of services provided by dozens of operators that includes fixed-route local bus, community circulators, express bus, bus rapid transit (BRT), demand response, commuter rail, heavy rail, and light rail. Ridership in our region continues to grow, and significant progress is being made in making transit more available and attractive by virtue of a burgeoning rail network, transit-oriented development (TOD), and other service improvements. Between

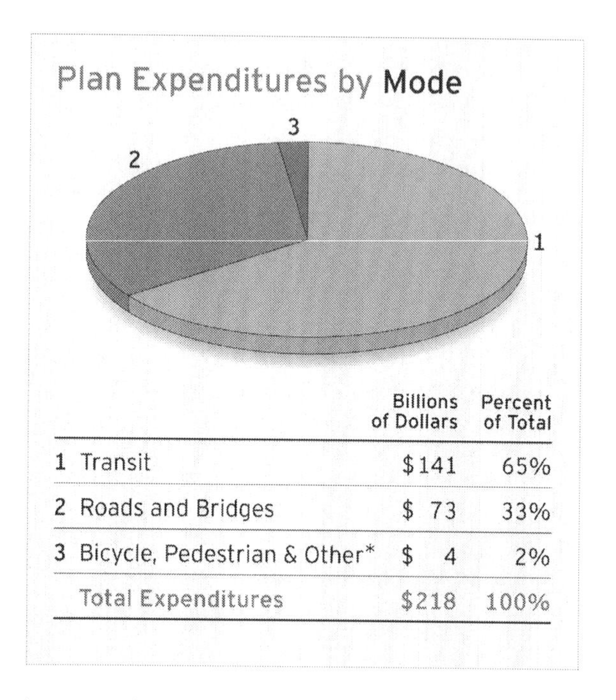

	Billions of Dollars	Percent of Total
1 Transit	$141	65%
2 Roads and Bridges	$ 73	33%
3 Bicycle, Pedestrian & Other*	$ 4	2%
Total Expenditures	$218	100%

Figure 6.39 Plan Expenditures by Mode.
(Metropolitan Transportation Commission, 2009, p. 37).

> 2000 and 2008, bus ridership increased by 17 percent, and urban rail ridership increased by 50 percent.
>
> (Southern California Association of Governments, 2012, p. 19)

These RTPs illustrate two other principles of the transit-oriented planning. First, a full range of transit service types should be provided, including the newest addition to the transit family, bus rapid transit. Second, land use changes should be planned for transit station areas in order to boost transit productivity.

On the range of services provided, Denver, Minneapolis-St. Paul, Portland, Salt Lake City, and others now have it all (Denver Regional Council of Governments, 2011; Minneapolis-St. Paul Metropolitan Council, 2010; Portland Metro Regional Government, 2010; Wasatch Front Regional Council, 2011). Take Salt Lake City:

> The 2040 RTP also recommends adding approximately 296 miles of major public transit improvements. These improvements include 12 additional miles of Light Rail Transit, 6 miles of Commuter Rail Transit, 161 additional miles of Bus Rapid Transit (BRT 3), 106 miles of

Enhanced Bus (BRT 1) service, and 11 miles of streetcar lines. These major transit improvements will provide an increase of over 60,000 revenue miles of transit service each weekday, or a 94 percent increase in current service. The 2040 RTP recommends that local bus route service throughout the Wasatch Front Region be increased by at least 25 percent over the next 30 years.

(Wasatch Front Regional Council, 2011, p. 18)

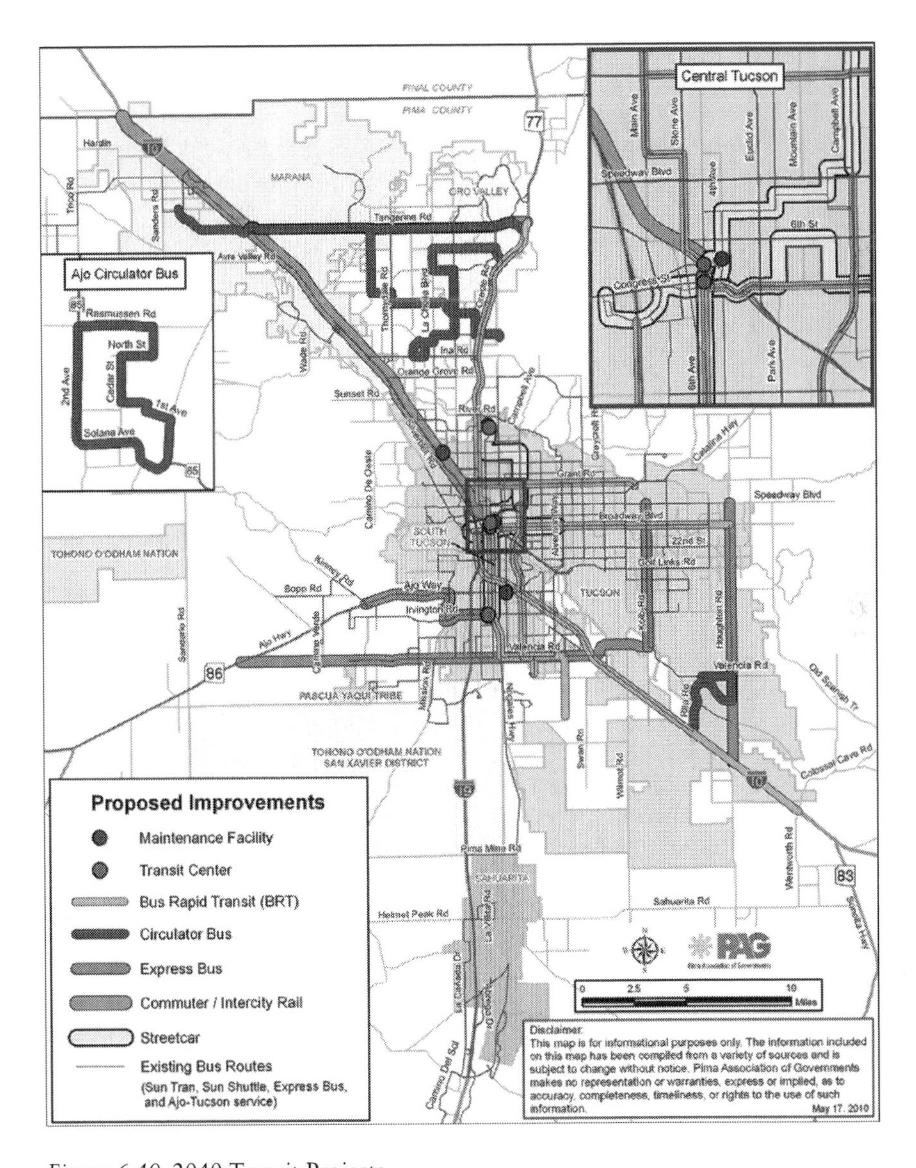

Figure 6.40 2040 Transit Projects.

(Pima Association of Governments, 2010, p. 36).

In smaller urbanized areas, "having it all" may mean having fewer than the full array of transit service types. In Tucson, for example, it means having new BRT and streetcar services and possibly commuter rail at some point, but no light rail (Pima Association of Governments, 2010, p. 37).

On the promotion of transit-oriented development around stations, many previously autocentric cities have followed the lead of Washington, DC, San Francisco, and Portland. Charlotte developed the 2025 Integrated Transit/Land Use Plan, redirecting development from auto-oriented wedges to transit-served corridors. It then took the plan to the voters, who approved a half-percent sales tax for transit in October 1998.

> Such transit investments include capital expansion projects like the LYNX Blue Line light rail extension to smaller projects, such as the replacement of bus engines. The primary funding source for these capital projects are a combination of federal and state grants in addition to local matching funds from the half-percent sales and use tax dedicated to transit within Mecklenburg County.
>
> (Charlotte Regional Transportation Planning Organization, 2014, p. 11–13)

Bicycle and pedestrian improvements

Background

Walking and bicycling are fundamental forms of locomotion (Pratt et al., 2012). All trips begin and end with walking, and short trips are ideal for both bicyclists and pedestrians. According to the 2009 National Household Travel Survey (NHTS), almost 11 percent of all trips in the United States are by walking, making it the second most popular mode after driving. One percent of all trips are by bicycle. Of trips a half mile or less, 62 percent are by walking and 3 percent are by bicycle. So Americans will walk or bike under the right conditions. The problem is that conditions are not "right" across most of the United States.

Walking and bicycling have implications for mobility, livability, social justice, and public health. The health benefits of walking and bicycling, in particular, are widely recognized. There are at least 14 surveys of the literature on the built environment and physical activity, including walking and biking (Badland & Schofield, 2005; Cunningham & Michael, 2004; Frank, 2000; Frank & Engelke, 2001; Humpel, Owen, & Leslie, 2002; Kahn, Ramsey, Brownson, Heath, & Howze, 2002; Krahnstoever-Davison & Lawson, 2006; Lee & Moudon, 2004; McCormack et al., 2004; National Research Council, 2005; Owen, Humpel, Leslie, Bauman, & Sallis, 2004; Saelens & Handy, 2008; Trost, Owen, Bauman, Sallis, & Brown, 2002. The literature is now so vast it has produced three reviews of the many reviews (Bauman & Bull, 2007; Ding & Gebel, 2012; Gebel, Bauman, & Petticrew, 2007).

Pedestrians and bicyclists are much slower than motorists, and travel without the benefit of a climate-controlled, entertainment-equipped, protective metal shell provided by a car. This makes them much more sensitive to the length of trips and the environment in which they travel. Three factors significantly affect rates of walking and bicycling: distance, safety, and design.

First, regions must be designed to keep trip distances short. The likelihood of walking and biking drops off dramatically after ½ mile and 2 miles, respectively (Litman, 2007; Pratt et al., 2012). Mixing uses in compact developments achieves this goal, particularly when combined with site planning that enhances direct nonmotorized travel throughout developments (Ewing, Greenwald, Zhang et al., 2011).

The substantial number of short trips undertaken every day in the United States indicates the potential for walking and biking to accommodate a greater percentage of daily trips. Even in the sprawled metropolitan areas of the United States, 49 percent of all trips are shorter than 3 miles, 40 percent are shorter than 2 miles, and 28 percent are shorter than 1 mile (Pucher & Dijsktra, 2000). Bicycling can easily cover all of these distances; however, Americans currently drive their cars for 89 percent of trips between 1 and 2 miles long (Pucher & Dijsktra, 2003). Not only are these distances ideal for bicycling and walking, they are also key distances for connecting residences to transit lines in many metropolitan regions. For this reason, transit agencies now prioritize pedestrian and bicycle infrastructure around transit stops, in First-Mile, Last-Mile initiatives.

Second, trips must be safe. This can be achieved through continuous networks with sidewalks, bicycle lanes, enhanced crossings, good lighting, and access control. Heath et al. (2006), for example, found a 51 percent increase in pedestrian volume after the installation of streetlights along poorly lit streets in London. Nearly all examples of "mass bicycle commuting" to school occur where access is possible by separate bicycle paths or by low-volume residential streets (Everett & Spencer, 1983).

A recent review by Pratt et al. (2012) compiled travel behavior studies from across the country, assessing the impact of bicycle and pedestrian facilities in particular. Investments ranged from sidewalks and street crossings to pedestrian/bicycle promotions. Simply providing sidewalks produced some of the greatest increases in use, with volumes increasing up to 400 percent in the compiled assessments (p. 14). Bicycle lane networks also led to large increases in cyclist traffic, with an average 51 percent increase over six studies (p. 22).

Finally, trips must be interesting. Interesting walks seem shorter than boring walks. Pedestrians use intermediate goals and destinations as points of orientation along the way, having the effect of psychologically shortening the sense of distance and time (Pratt et al., 2012). This effect can be achieved through good urban design (Ewing & Handy, 2009). On pedestrian streets, you design for *imageability*, *visual enclosure*, *human scale*, *transparency*, and *complexity*. Three research efforts have tested measures of aesthetics and found 25 and even 41 percent positive differentials in pedestrian and bicycle activity in prescribed circumstances (Parsons Brinckerhoff, 1996; Heath et al., 2006; Committee on Physical Activity, Health, Transportation, and Land Use, 2005, p. 118).

Regarding walking, physical features, urban design qualities, and individual reactions may all influence the way an individual feels about the environment as a place to walk (Ewing & Handy, 2009). In two related studies, pedestrian counts on hundreds of street segments were found to be related not only to gross qualities such as population density and block size, but also to micro characteristics of streetscapes such as windows on the street and buildings lined up to form "streetwalls" (Ewing & Clemente, 2013; Ameli et al., 2015). Indeed, in one study, these micro characteristics of the street environment proved to be the most significant influences on pedestrian traffic (Ewing & Clemente, 2013).

Regarding bicycling, existing and potential bicyclists can be categorized into a few distinct groups. They estimate that 1 percent of the population is strong and fearless when bicycling and 7 percent is enthused and confident regarding bicycling, whereas 60 percent is interested but concerned regarding cycling and the remainder will never bicycle regularly, "no way, no how" (Geller, 2009; Dill, 2012). These categories provide some indication for how to target investments to encourage bicycling.

According to this model, places that have achieved bicycle mode shares over 30 percent have started to successfully accommodate this "interested but concerned" demographic. Copenhagen, which vies with Amsterdam for top bicycling city in the world with a mode share approaching 40 percent, sought to determine the fundamental reasons why such a large proportion of their residents choose bicycling over other modes. In comprehensive surveys of residents, the Copenhagen Bicycle Office determined that the top reasons people choose bicycling are that it is fast, according to 55 percent of respondents, and convenient, according to 33 percent of respondents. Meanwhile, only 5 percent of Copenhageners report that they feel "very unsafe" when bicycling in the city (Copenhagen Bicycle Office, 2012).

The presence of bicycle facilities has been consistently correlated with higher rates of cycling (Nelson & Allen, 1997; Macbeth, 1999; Dill & Carr, 2003; Buehler, Hamre, Sonenklar, & Goger, 2009; Duthie, Brady, Mills, & Machemehl, 2009; Queensland Transport, 2007; Buehler & Pucher, 2012; Parker et al., 2013). American cities and towns that have installed comprehensive off-street and on-street bicycle networks are now accommodating a large proportion of these daily trips by bicycle: in Portland, Oregon, 7 percent of work trips are accommodated by bicycle, whereas 9 percent of trips are by bicycle in Boulder, Colorado, and 22 percent of trips are by bicycle in Davis, California. These cities have pursued intense expansions of bikeways; in Boulder, 95 percent of arterials have adjacent bike lanes or adjacent pathways (Pucher & Buehler, 2012).

With a sample of 18 cities, Nelson and Allen (1997) determined that every additional mile of bicycle paths and lanes per 100,000 residents corresponded to a 0.069 percent increase in the number of people bicycling to work. Dill and Carr (2003) collected a sample of 42 cities, and found that after controlling for factors such as weather and spending on infrastructure, for every additional mile of bicycle paths and lanes, bicycle ridership increased 1 percentage point.

In terms of facility type, there is a clear preference for shared-use bicycle paths and separated bicycle lanes, also called cycle-tracks, over riding in lanes shared with auto traffic, and people will bicycle out of their way in order to use bicycle paths and bicycle lanes (Barnes, Thompson, & Krizek, 2006; Moudon et al., 2005; Krizek, Barnes, et al, 2007; Dill, 2009; Lusk et al., 2011). In a larger sample of 92 cities, Buehler and Pucher (2012) controlled for a variety of demographic, weather, and built environment characteristics and found that a 10 percent greater supply of bike lanes is associated with a 3.1 percent increase in the level of bicycle commuting in a city, whereas a 10 percent greater supply of bicycle paths is associated with a 2.5 percent increase in the level of bicycle commuting.

Characteristics of European-style bicycle networks that achieve speed, convenience, and comfort are beginning to appear in the United States. These bikeways separate bicycles from auto traffic on arterials and collectors, prioritize bicycle traffic at intersections, and create low-speed, low-volume environments on minor streets that are shared with automobiles. Cycle-track designs that provide a physical separation between auto traffic lanes and bicycle lanes have been shown to increase both safety and ridership over streets without facilities and streets with conventional striped bike lanes (Lusk et al., 2011; Chicago Department of Transportation, 2012; San Francisco Municipal Transportation Authority, 2011; District Department of Transportation, 2011). Similar results have been achieved by bicycle boulevards, in which traffic diversions and traffic calming on low-volume, low-speed, high-connectivity neighborhood streets combine with signage and pavement markings to prioritize bicycle traffic over auto through-traffic (Portland Bureau of Transportation, 2012).

Regarding design standards, cities are pioneering new guidelines that emphasize separation and protection of bicycles in an effort to encourage the interested but concerned demographic. The National Association of City Transportation Officials (NACTO), a coalition of transportation professionals from major cities throughout the United States, has released two editions of a manual that incorporates separated facilities such as cycle-tracks, bicycle-priority designs such as bicycle boulevards, and traditional bicycle lane designs (NACTO, 2012). In contrast to the new standards adopted by major U.S. cities, the latest national-level guidance, the *AASHTO Guide for the Planning, Design and Operation of Bicycle Facilities*, focuses solely on conventional bike lane designs adjacent to automobile traffic (AASHTO, 2012).

Conventional practice

The conventional practice is to treat bicycle and pedestrian travel as somewhat incidental to the regional transportation system. It is mentioned in the RTP, in a paragraph or section, but is in no way a priority of the plan: "pedestrian issues have not been embodied in MARC's previous transportation plans in the past other than in a cursory fashion" (Mid-America Regional Council, 2005, p. 7–10).

> While many of the denser parts of the region have comprehensive sidewalk networks, the more rural and recently developed suburban areas have been designed primarily for the automobile, as pedestrian facilities such as sidewalks and crosswalks are not consistently included in roadway projects and many intersection designs include free-flowing turn lanes. Even in areas with comprehensive sidewalk networks, there are still significant needs.
>
> (Baltimore Regional Transportation Board, 2007, p. 45)

In many regions, bicycle and pedestrian travel is not forecasted along with auto and transit travel. The region (or state within which it is located) has no "complete streets" policy to guarantee that new roadways accommodate bicycle and pedestrian users. And the region has no plan for closing gaps in the existing bicycle and pedestrian networks.

Best practice

"Measures that lead to a more sustainable life-style are comprised of strategies that reduce congestion, increase access to public transportation, improve air quality, and enhance coordination between land use and transportation decisions" (Oahu Metropolitan Planning Organization, 2011, p. 31). Between 1991 and 2004, the City of Portland invested $12 million in the city's bikeway network, increasing the mileage from 78 to 256. The network includes bike lanes and designated "bike boulevards" – low-traffic city streets suitable for bicycling. Bicycle counts in 2012 showed a 128 percent increase in bicycle traffic across the bridges over the past 10 years (Portland Metro Regional Government, 2014, p. 1–54).

One best practice with respect to bicycle and pedestrian travel is to adopt ambitious mode share targets. The Seattle MPO has set a goal of 20 percent of all trips by biking and walking by 2030 (Puget Sound Regional Council, 2007, p. iv). The previous mode share was 5 percent (Puget Sound

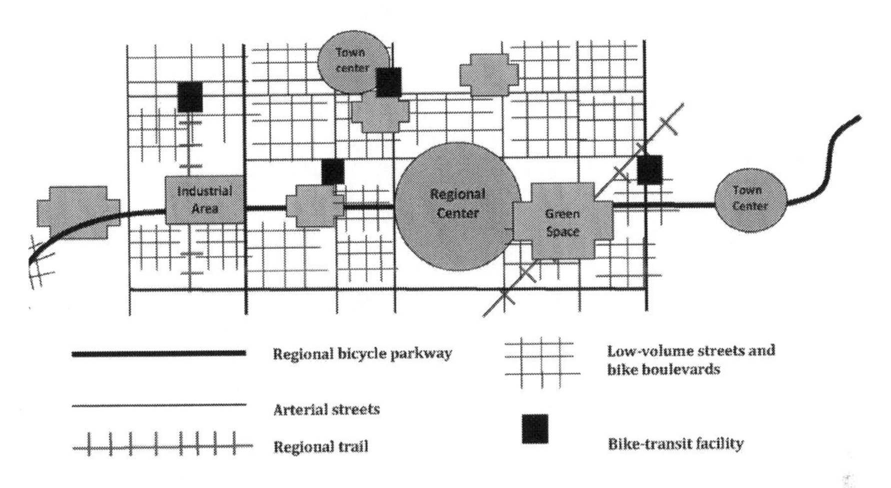

Figure 6.41 Regional Bicycle Network Vision for Portland, OR.
(Portland Metro Regional Government, 2014, p. 2–66).

Regional Council, 2007, p. 15). The current RTP has walk and bike trips increasing 50 percent faster than population (Puget Sound Regional Council, 2010, p. 38). The Dallas-Ft. Worth MPO has set a goal of 8 percent of all trips by biking and walking by 2030 (North Central Texas Council of Governments, 2009, p. 172). The mode share in the 1996 household travel survey was 5.5 percent (North Central Texas Council of Governments, 2009, p. 171).

Another best practice is to adopt a complete streets policy so all future roadway projects accommodate bicyclists and pedestrians. Some MPOs, like New York's, have no formal policy but rely on state and local governments to pursue complete streets:

> The region is at the forefront of designing and operating transportation infrastructure that supports all types of travel. NYMTC members continue to develop what have been called "complete streets," streets that are open and safe for all users. Benefits include improved access to the transit system, which encourages higher transit ridership and discourages auto use.
>
> (New York Metropolitan Transportation Council, 2010, p. 1–29)

In 2011, New York State passed Complete Streets, which requires NYMTC agencies to consider complete streets design principles on all future projects which receive either federal or state funding (New York Metropolitan Transportation Council, 2010, p. 4–29).

Other MPOs, like St. Louis', try to cajole constituent governments into providing complete streets:

> In 2006, the Council launched the Great Streets Initiative to expand the way communities think of transportation. Rather than viewing a roadway project as solely a way to move more cars and trucks faster, the goal of the St. Louis Great Streets Initiative is to trigger economic and social benefits by centering communities around interesting, lively and attractive streets that serve all modes of transportation.
>
> (East-West Gateway Council of Governments, 2007, p. 85)

By the time of the current RTP, in 2013, the MPO was able to report:

> Communities around the St. Louis region are joining in the Complete Streets Movement to make their streets safe, accessible and healthy. In total, seven municipalities in the region have adopted Complete Street policies, namely the cities of St. Louis, Ferguson, Crystal City, De Soto, Festus, Herculaneum and Pevely.
>
> (East-West Gateway Council of Governments, 2011, p. 38)

Going a step further, the Sacramento MPO, in coordination with the local Complete Streets Coalition, has developed a Complete Streets Resource Toolkit. The toolkit is part of SACOG's complete streets technical assistance program. The toolkit includes such things as fact sheets, case studies, presentations, and photo simulations. It puts a wealth of resources at the fingertips of any advocate, community member, planner, or engineer (Sacramento Area Council of Governments, 2012, p. 60).

A third best practice involves the retrofitting of existing streets with sidewalks and bike lanes. Sidewalks are currently provided on about 70 percent of the regional roadway system arterials within the Denver urban growth boundary. An additional 500 linear miles are needed to complete the system (Denver Regional Council of Governments, 2011, p. 64). Even the Portland region has major gaps in its pedestrian network. In 2001, the region had 1,230 miles of potential pedestrian facilities in transit/mixed-use corridors and pedestrian districts. However, only 821 miles of those 1,230 potential miles had sidewalks, for a pedestrian system that was only 66 percent complete (Portland Metro Regional Government, 2010, p. 1–56). Sidewalk data for the region collected in 2011 indicates that 38 percent of regional pedestrian corridors are missing sidewalks on at least one side of the roadway (Portland Metro Regional Government, 2014, p. 159).

In this regard, the best practice is to develop bicycle and pedestrian master plans for completing these networks, and to fully fund these plans. Atlanta,

Denver, and St. Louis have adopted such plans (Atlanta Regional Commission, 2014; Denver Regional Council of Governments, 2011; East-West Gateway Council of Governments, 2011). Each New York sub-area has developed its own pedestrian and bicycle plans to guide future investments in nonmotorized transportation (New York Metropolitan Transportation Council, 2010, p. 6–21). San Diego has developed a regional bicycle plan, and the San Diego MPO requires local agencies to develop pedestrian master plans in order to be eligible for discretionary, nonmotorized funding administered by the MPO (San Diego Association of Governments, 2007, p. 6–50;

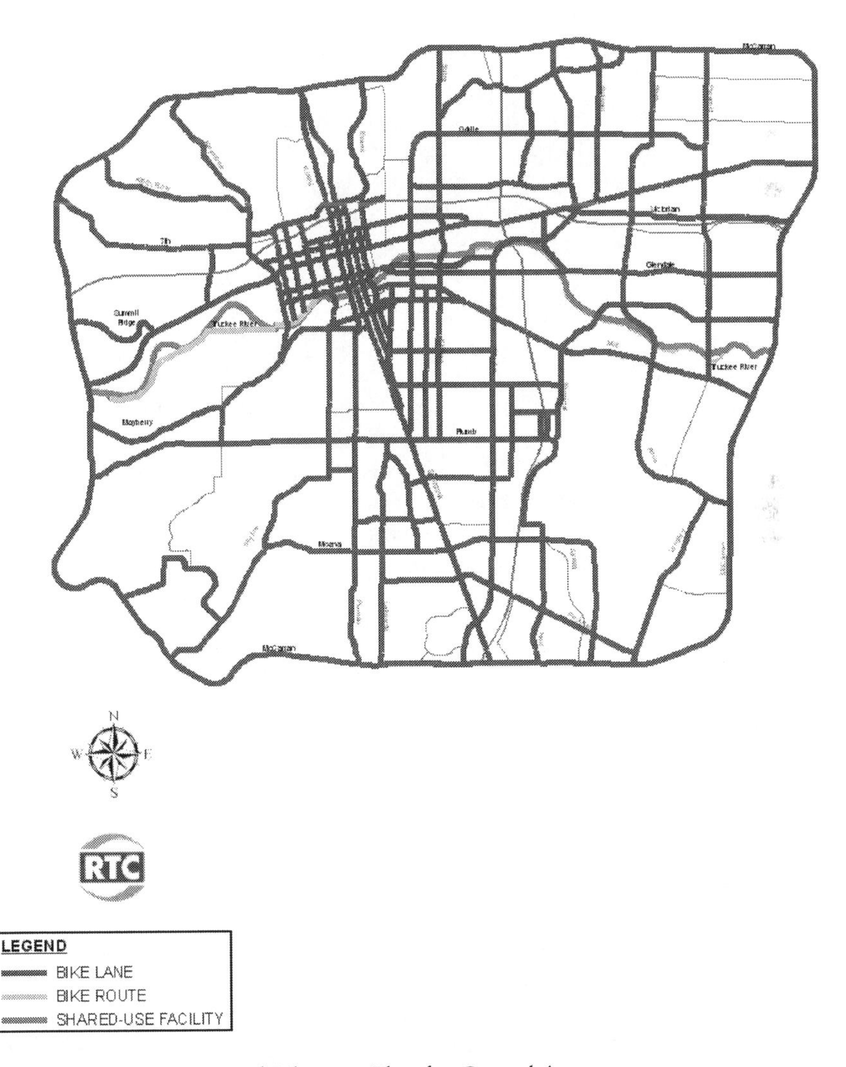

Figure 6.42 Reno Regional Bikeways Plan for Central Area.
(Regional Transportation Commission of Washoe County, 2008, p. 5–7).

San Diego Association of Governments, 2011, p. 6–53). The Honolulu MPO has incorporated projects from the Bike Plan Hawaii and is developing a pedestrian master plan (Oahu Metropolitan Planning Organization, 2006, p. 7). The latter is not mentioned in the more recent RTP (Oahu Metropolitan Planning Organization, 2011). The Reno MPO has a Regional Bikeways Plan that will place bike lanes on nearly all roadways in the central area of the region. The RTP anticipates that 80 percent of the plan will be completed by 2020, and that 100 percent will be completed by 2040 (Regional Transportation Commission of Washoe County, 2008, p. 2–10).

Of those regions earmarking funds for bicycle and pedestrian facilities, none spends as much on those facilities as the mode share would seem to justify. In Sacramento, 8 percent of total funds ($4.0 billion in escalated costs) are earmarked for exclusive bicycle and pedestrian improvements, including bicycle trails, sidewalks, ADA retrofits, and supporting facilities (Sacramento Area Council of Governments, 2012, p. 56). In addition, 25 percent of the road capital projects have a bicycle or pedestrian feature that is not included in the total (Sacramento Area Council of Governments, 2008, p. 9). That is one of the highest percentages among featured RTPs, but still less than the bike-pedestrian mode share.

The Denver MPO envisions that by 2035 sidewalks or multipurpose trails will be provided along all applicable roadways within the urban growth boundary (Denver Regional Council of Governments, 2011, p. 61). The San Francisco MPO has earmarked $1 billion to fully fund and complete its bikeway network (Metropolitan Transportation Commission, 2009, p. 16).

A fourth best practice is to forecast bicycle and pedestrian travel as available mode choices, accounting for mode shifts as facilities are improved and land use patterns become more compact. The common failure to even acknowledge the possibility of nonmotorized trips puts these modes at a competitive disadvantage vis-à-vis motorized modes when it comes to funding decisions. Enhancing four-step travel demand models to include nonmotorized trips in trip generation and mode choice steps, or using post-processing to split trip tables between motorized and nonmotorized modes, represent best modeling practices (see Transportation and Land Use Modeling).

Project selection

BackgroundMAP-21 transformed the Federal-aid highway program and the Federal transit program by requiring a transition to performance-driven, outcome-based approaches to key areas. With respect to planning, the statute introduced critical changes to the planning process by requiring States, MPOs, and operators of public transportation to link investment priorities to the achievement of performance targets in key areas such as safety, infrastructure condition, congestion, system reliability, emissions, and freight movement. With respect to planning, the FAST Act (2015 successor to MAP

-21) left the provisions of MAP-21 intact and made minor revisions to existing provisions.

In selecting projects for funding and construction, MPOs commonlyrefer to federal legislation and regional transportation planning criteria for guidance. The degree to which different MPOs adhere to the guiding principles of federal guidance and their own planning criteria, however, varies widely. In literature on this topic, this variance relates to:

- The degree to which scoring criteria accurately reflect state, regional, and local goals,
- The strictness and comprehensiveness of project scoring criteria, and
- The quality of performance measures tied to these criteria.

In a review of regional long-range transportation plans, also known as Metropolitan Transportation Plans (MTPs), the U.S. Department of Transportation Volpe Center performed seven case studies of model MTPs to determine best practices for a range of plan elements, including project selection (Lau, 2012). The review found that the best MTPs not only provided overall direction for project selection, but created explicit links to corridor or area studies and plans, and connections between project selection and screening criteria based on goals laid out in the MTP. Several of the model MTPs in this study accomplished this connection through strategies such as eligibility formulas, points systems, or checklists referring directly to MTP goals. In this way, long-range planning goals were able to direct federal and other fund allocations. A successful example of one such mechanism is the Transportation Implementation Plan scoring formula used in the Portland Area Comprehensive Transportation System (PACTS), which ranks projects based on eight long-range goals established in the MTP. This score in turn accounts for half of the selection score for each project (Lau, 2012).

The most successful MTPs created a clear link between over-arching goals, specific objectives, and performance measures. The best examples established measures that were meaningful to partners such as transit agencies, local governments, and the general public (Lau, 2012). For example, the PACTS model used Vehicle Hours Traveled, which helps partners and the general public understand the time cost of congestion.

Despite the development of such performance measures, many regions, including New York City, struggled to incorporate SAFETEA-LU goals such as resiliency and sustainability into project selection (Paaswell, 2013). In an attempt to address these gaps, the Lexington (Kentucky) area MPO undertook an effort to formalize and enhance the project selection process for their MTP by reviewing processes used by other MPOs and incorporating each of SAFETEA-LU's planning factors, as well as several goals unique to the Lexington MPO. The final list of scoring criteria included 18 different elements, including economic impact, multimodal involvement, and public support (Schaub, 2010). The new criteria were then used to score potential projects for incorporation into the new MTP, analyzing projects from the previous regional transportation plan, from congestion and corridor studies, and from comprehensive plans in the region. After scoring each project, MPO staff provided an additional qualitative score, developed a cost estimate, and then ranked the projects in light of a fiscal constraint analysis. MPO staff reported satisfaction with this approach (Schaub, 2010).

More recently, Moving Ahead for Progress in the 21st Century (MAP-21) increased the emphasis on sustainability and resiliency elements by establishing livability, equity, and environment as planning factors, helping spur the development of additional scoring criteria and performance measures, such as a Transportation Cost Index (TCI). This measure, under development in Oregon, would create a method for consistently comparing multimodal performance across geographic areas, including state, MPO, and local levels of decision making (Wang, Liu, & Yang, 2015).

The Volpe Center report notes that the best MTPs also influence project selection more indirectly by creating feedback loops between regional plans, local plans, modal plans, and corridor or area studies, such as a local transit or pedestrian and bicycle plan (Lau, 2012). Model MTPs call for new local or small-area studies to guide development of future local projects. In turn, if a project is prioritized in a local plan or study, the project receives a higher ranking in the regional project selection scoring process. This feedback loop increases in complexity when technical committees develop an additional layer of scoring criteria based on goals articulated in small-area or modal studies. Eventually, the MPO can incorporate these local or modal principles back into the MTP and subsequent project selection processes, enhancing the relevance of the regional plan for local entities.

Despite this coordination, additional funding sources with separate award cycles may disrupt the link between local project selection and regional priorities. For example, the Transportation Investments Generating Economic Recovery (TIGER) program awards significant project funding directly to local entities, and despite federal mandates for coordination with MPOs, many regional councils report no involvement with the process (Lowe, 2014). Furthermore, travel-sheds often stretch beyond regional government boundaries and depend on land use patterns outside the scope of a given MPO, reducing effectiveness in addressing local and regional travel patterns. Given this situation, there is increasing emphasis on developing broader institutional coordination and alternative arrangements among government entities to increase agreement on desired outcomes related to transportation decisions and project selection (Margerum, 2010).

By clearly establishing the link between regional plan goals and project selection, model MTPs increase the importance of the regional plan for local communities and community groups (Lau, 2012). In this way, communities have an incentive to create local plans and subsequent projects that support regional goals. Furthermore, this consistent and coherent interrelationship between MTP goals and project selection increases the investment of stakeholders in development of regional goals that reflect local needs and priorities.

Conventional practice

The conventional practice is to base funding decisions primarily on roadway levels of service or their equivalent. Travel demand models are used to forecast future traffic flows on different links in the existing and committed roadway network. Links whose future traffic volumes exceed capacity at the adopted roadway level-of-service standards then receive funding priority. Travel demand models are also used to test if links in the future roadway network, with capacity improvements, will operate at acceptable levels of service.

Whereas the conventional practice has been tempered in most metropolitan areas, vestiges of the old systems remain. Many MPOs still use roadway level-of-service measures and standards to determine deficiencies in roadway networks. Several still prioritize projects on the basis of these determinations. If anything, this practice is encouraged by federal requirements that large regions have congestion management processes.

The 2009 Chattanooga RTP, for example, estimates levels of service based on the following criteria:

$V/C \leq 0.60$ – LOS A
$0.61 < V/C \leq 0.70$ – LOS B
$0.71 < V/C \leq 0.80$ – LOS C
$0.81 < V/C \leq 0.90$ – LOS D
$0.91 < V/C \leq 1.0$ – LOS E
$V/C > 1.0$ – LOS F

V/C is the volume-to-capacity ratio on a given link. The RTP uses LOS D as an acceptable level of service. Less than LOS D indicates unacceptable congestion, constitutes a failure in LOS terms, and entitles a link to funding priority (Chattanooga Hamilton County North Georgia Transportation Planning Organization, 2009, p. 3–14).

The Lansing RTP defines links with V/C ratios of 1.0 or greater as deficient and those with V/C ratios of 0.8 to 1.0 as near-deficient. Both receive funding priority:

> link performance begins to degrade well before the V/C ratio reaches 1.0. Links with a V/C ratio greater than 0.8 were identified as near-deficient and are also considered congested, since they are approaching capacity. Near deficient locations need to be monitored for worsening conditions. “Near deficient” links may be candidates for operational or other remedial treatments to prevent them from becoming deficient . . . Access management, mandatory traffic impact studies for new land

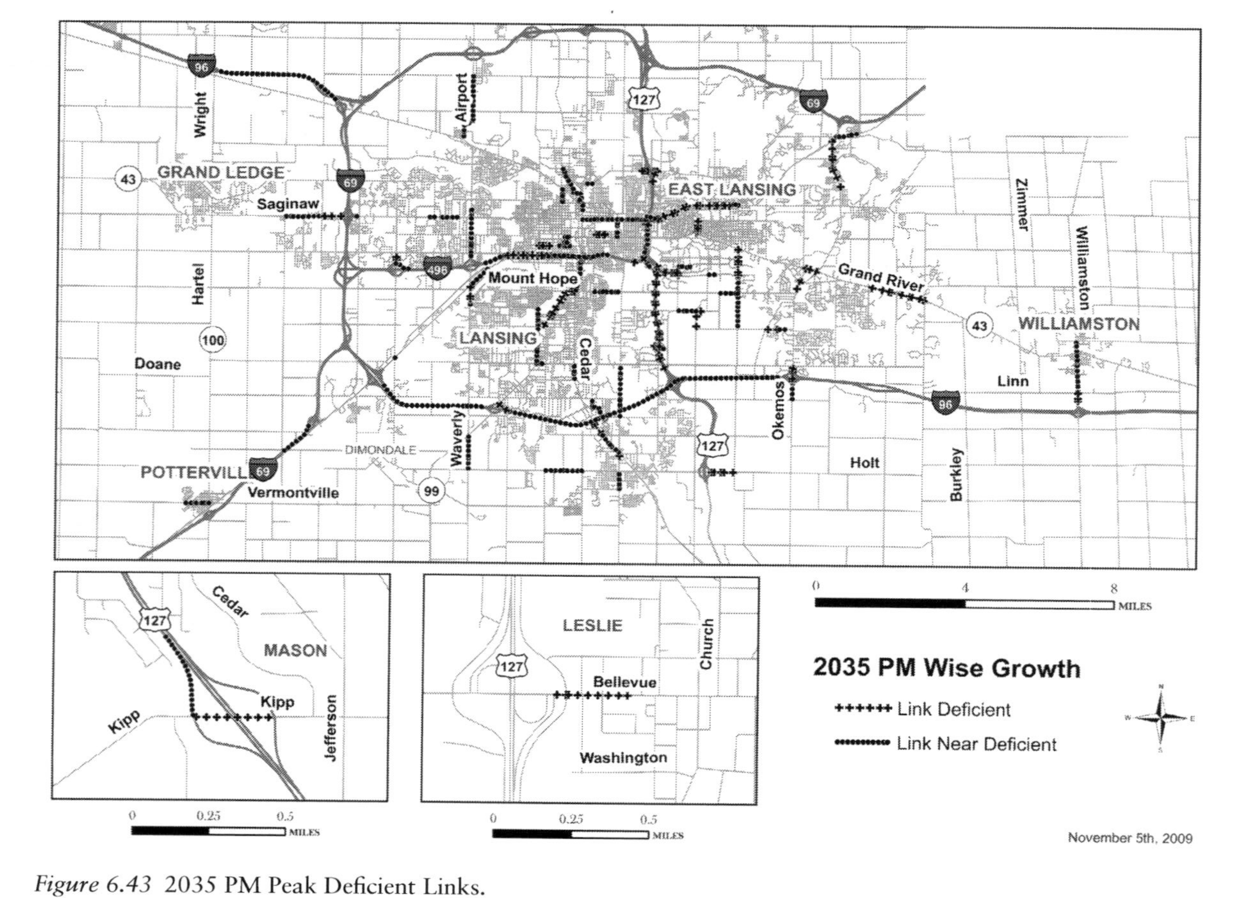

Figure 6.43 2035 PM Peak Deficient Links.
(Tri-County Regional Planning Commission, 2010, p. 10–15).

development or other techniques should be considered to prevent further degradation of these facilities.

(Tri-County Regional Planning Commission, 2010, p. 10–10)

Best practice

The best practice is to recognize that regions "cannot pave their way out of congestion," and that thriving downtowns and other activity centers are bound to have some congestion. The best practice is to prioritize funding based on broad public purposes. The inability for a region to build roads fast enough to keep up with demand is acknowledged in so many words by the Baltimore, Cincinnati, Eugene, Raleigh-Durham, and Tucson MPOs.

> The RTP recognizes that sole reliance on more and bigger roadways to meet the transportation demand is shortsighted. Even if adequate funding was available, given the growth anticipated in the region, it is unreasonable to assume the region can build its way out of traffic congestion. The technical evaluation of TransPlan alternatives indicated that the travel demand associated with growth will overload the transportation system, even with major capacity-increasing projects. Experience from cities all over the world suggests that building roads encourages more people to use cars, thereby perpetuating the transportation challenges.
>
> (Central Lane Metropolitan Planning Organization, 2007, Chapter 1, p. 4)

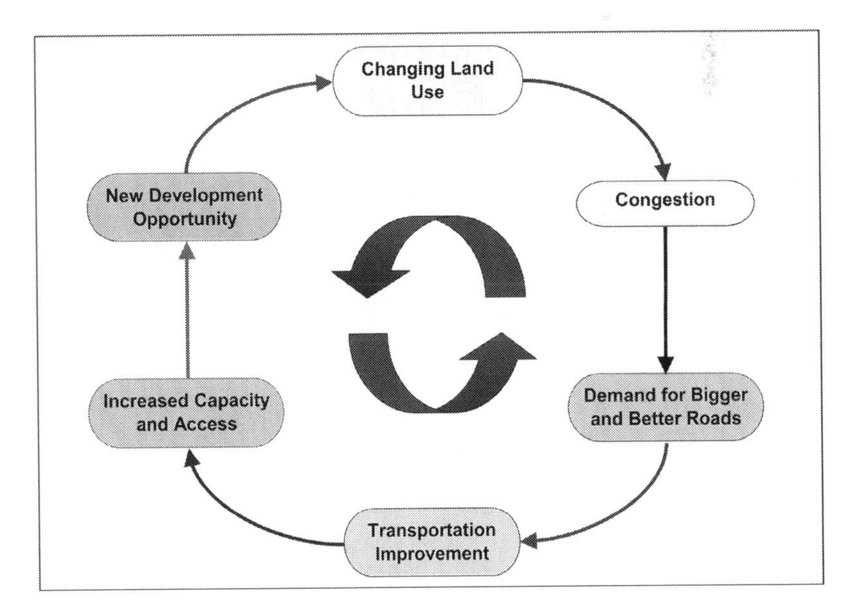

Figure 6.44 The Land Use-Transportation Cycle.
(Tri-County Regional Planning Commission, 2010, p. 2–68).

We have already argued that as a best practice, the following project categories deserve funding priority: maintenance projects under a fix-it-first policy, TSM and TDM projects under a cost-effectiveness policy, and new capacity projects serving existing centers under a smart growth policy.

For other new capacity projects, funding criteria can be based on the broad goals and objectives that nearly all MPOs have adopted. They can also be based on the federal metropolitan planning factors whose consideration has been mandated in federal legislation since ISTEA.

The 2007 Seattle RTP prioritizes as follows:

1 The first priority should be to maintain, preserve, make safe and secure, and optimize existing transportation infrastructure and services. The most cost-effective infrastructure investments are usually those that maintain, preserve, improve safety and security, and optimize existing assets.
2 Investments should emphasize continuity and complete discrete elements of the transportation system. Completing missing pieces of larger systems is a regional investment priority.
3 Appropriate investments in all modes should be emphasized to provide an array of travel choices.
4 Transportation investments should be directly linked with measurable transportation, environmental and land use outcomes, and should support the achievement of regional and state benchmarks.
5 Compact development of designated urban centers, high capacity transit station areas, and other communities should be supported through direct investment. Projects that serve and support greater concentrations of activity within the Urban Growth Area are also regional priorities.

(Puget Sound Regional Council, 2007, p. 22)

For highway capacity projects, project scoring criteria have been linked to broad public goals by MPOs in Baltimore, Boston, Charlotte, Cincinnati, Denver, Indianapolis, and St. Louis, among others. Typically, MPO staff score competing projects on a Likert scale with varying points (0 to 2, 1 to 5, etc.) assigned to each of the criteria. Scores are then summed and used to prepare a list of projects in priority order. The staff list is submitted to the MPO's board of directors for approval. The voting members of the board have the ultimate say over what is funded, up to the limit of available revenues.

One legitimate concern about the project scoring methodology described above is the subjective and staff-driven nature of the process. In the interest of transparency, it is desirable that scoring criteria be as objective as possible. Tallahassee has done a particularly good job of operationalizing its funding criteria. The criteria themselves are based on the RTP's goals and objectives, which themselves are based on federal metropolitan planning factors, input from state and local officials, and suggestions from the public. Scoring criteria relate to each goal. The points awarded for each criterion are about as objective as one can make them. Some examples:

Existing capacity deficiency

Presently no documented capacity problem 0
Presently at or near capacity (100 or less trips available) 1
Presently over capacity (negative trips available) 2

Promote sustainable development score

Project could promote urban sprawl 0
Project recommended as part of a Sector/Master Plan outside urban core 1
Project could promote urban infill or transit-oriented development 2

Hurricane evacuation

Project not likely to enhance hurricane evacuation 0
Project should enhance hurricane evacuation 1

Title VI impacts

Project has no positive impact to/from/within Title VI areas 0
Project may improve accessibility to/from/within Title VI areas 1
Project likely to improve economic opportunities to/from/within Title VI areas 2.

(Capital Region Transportation Planning Agency, 2006, p. 3–1)

Additional criteria and associated scores were added to the 2010 Tallahassee RTP (Capital Region Transportation Planning Agency, 2010, pp. 73–76).

Funding gaps

Background

Federal regulations require MPOs to produce "financially constrained" long-range transportation plans, where project expenditures stay within "reasonably expected" revenues. Many MPOs also conduct needs assessments and produce needs-based plans. Needs typically exceed reasonably expected revenues by a third or more.

As one RTP put it:

> Federal and state funding levels are not expected to increase, and the region's local funding contribution is lower than other large metropolitan areas across the country. Recognizing these issues, DVRPC

> has formulated a list of local funding options that could be used to finance improvements to the region's transportation system. During the extensive public outreach conducted as part of the development of Connections, DVRPC outlined the challenge to increase local funding for transportation investments.... The Connections Plan does not advocate any particular local funding alternative, but instead issues a challenge to the region's leaders, stakeholders, and citizenry to reach consensus on new local and regional means to maintain and modernize the region's critical transportation infrastructure, which impacts both our standard of living and our economic competitiveness. Given the large set of needs that will remain unmet at currently available funding levels, the region needs to seek ways to close its funding gap.
>
> (DVRPC, 2009. p. 2)

Conventional practice

The San Diego MPO has $30 billion in needs beyond its $48 billion financially constrained plan (San Diego Association of Governments, 2011, p. 5–14). The Dallas-Ft. Worth MPO estimates that $98.7 billion is needed to through 2035, but $393.5 billion would be required to "eliminate the worst levels of congestion" (North Central Texas Council of Governments, 2013, p. 2.1). The Atlanta MPO has $64.4 billion in needs beyond its $58.6 billion financially constrained plan (Atlanta Regional Commission, 2014, p. 4–2).

Faced with these huge funding gaps, MPOs typically throw up their hands and accept the shortfall. In Portland:

> Forecasted revenues, in the financially constrained plan, available for local OMP expenditures fall short of this ideal level of OMP revenues,

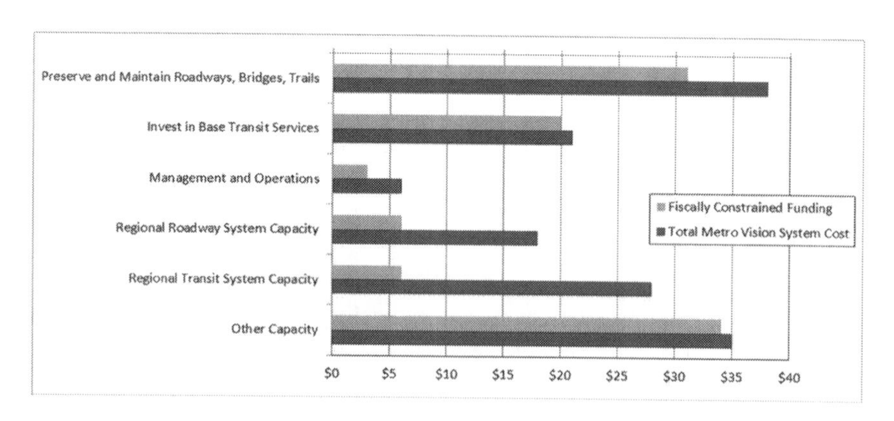

Figure 6.45 2040 Unconstrained Costs and Fiscally Constrained Revenues by Expense Category.

(Denver Regional Council of Governments, 2017, p. 91).

> which range from approximately $186 million in 2010 to $513 million in 2040; roughly 70 percent of "ideal" levels. However, this level of investment is fairly steady and represents the level of OMP investment in the regional street system that maintains the system at current conditions. While not ideal, this level of investment meets federal guidelines.
> (Portland Metro Regional Government, 2014, p. 3–29)

Best practice

The best practice is to aggressively pursue new and innovative sources to make up the funding shortfall. The Denver MPO outlines the following Transportation Funding Action Strategies in its RTP:

- Encourage the provision of local and private sector funds for use on transportation facilities that primarily serve local and private development access needs;
- Continue to ensure the region receives an equitable distribution of federal and state transportation funds;
- Support local, regional, and state efforts to increase transportation revenues necessary to meet the region's transportation needs, including tolling initiatives as appropriate;
- Actively seek federal discretionary funding for regionally significant transportation projects;
- Promote cooperation among elected officials, the business community, citizen groups, the Colorado Department of Transportation, and the Regional Transportation District in seeking new funding sources.
- Develop opportunities for implementing congestion pricing and other tolling techniques on existing freeways, and implement a tolling component (price-management) on new freeway lane-addition projects, where feasible, with all impacted communities included in the tolling decision and surplus revenue directed to multimodal investment or system preservation; and
- Support legislation that would implement VMT-based fees, pay-as-you-drive insurance, and other pricing strategies that more directly and immediately reflect the cost of vehicle travel to the user.

(Denver Regional Council of Governments, 2011, p. 110)

The Southern California Association of Governments also assumes a substantial amount of new funding:

> The plan includes a core revenue forecast of existing local, state, and federal sources, along with reasonably available new revenues sources that are likely to materialize within the RTP time frame. These new sources include adjustments to state and federal gas tax rates based

> on historical trends and recommendations from two national commissions (National Surface Transportation Policy and Revenue Study Commission and National Surface Transportation Infrastructure Financing Commission) created by Congress, further leveraging of existing local sales tax measures, value capture strategies, potential national freight program/freight fees, as well as passenger and commercial vehicle tolls for specific facilities. Reasonably available revenues also include innovative financing strategies, such as private equity participation.
>
> (Southern California Association of Governments, 2012, p. 7)

With a total budget of $524.7 billion over the next 25 years, $219.5 billion in funding is projected to come from new sources and innovative financing strategies.

Table 6.4 New Sources of Funding and Innovative Financing Strategies

Revenue Source	*Description*	*Amount*
Bond Proceeds from Local Sales Tax Measures	Issuance of debt against existing sales tax revenues: Los Angeles, Orange, Riverside, and San Bernardino Counties.	$25.6 bil
State and Federal Gas Excise Tax Adjustment to Maintain Historical Purchasing Power	Additional $0.15 per gallon gasoline tax imposed at the state and federal levels starting in 2017 to 2024 – to maintain purchasing power.	$16.9 bil
Mileage-Based User Fee (or equivalent fuel tax adjustment)	Mileage-based user fees would be implemented to replace gas taxes – estimated at about $0.05 (in 2011 dollars) per mile starting in 2025 and indexed to maintain purchasing power.	$110.3 bil (est. increment only)
Highway Tolls (includes toll revenue bond proceeds)	Toll revenues generated from SR-710 North Extension, 1–710 South Freight Corridor, East-West Freight Corridor, segment of the High Desert Corridor, and Regional Express/ HOT Lane Network.	$22.3 bil
Private Equity Participation	Private equity share as may be applicable for key initiatives: e.g., toll facilities; also, freight rail package assumes railroads' share of costs for main line capacity and intermodal facilities.	$2.7 bil
Freight Fee/National Freight Program	A national freight program is anticipated with the next federal reauthorization of the surface transportation act. The U.S. Senate's proposal would establish federal formula funding for the national freight network.	$4.2 bil

Revenue Source	*Description*	*Amount*
E-Commerce Tax	Although these are existing revenue sources, they generally have not been collected. Potentially, the revenue could be used for transportation purposes, given the relationship between e-commerce and the delivery of goods to California purchasers.	$3.1 bil
Interest Earnings	Interest earnings from toll bond proceeds.	$0.2 bil
State Bond Proceeds, Federal Grants & Other for California High-Speed Rail Program	State general obligation bonds authorized under the Bond Act approved by California voters as Proposition 1A in 2008; federal grants authorized under American Recovery and Reinvestment Act and High-Speed Intercity Passenger Rail Program; potential use of qualified tax credit bonds; and private sources.	$33.0 bil
Value Capture Strategies	Assumes formation of special districts including use of tax increment financing for specific initiatives.	$1.2 bil

(Southern California Association of Governments, 2012, p. 7)

Air quality conformity

Background

The 1990 Clean Air Act Amendments (CAAA) require MPOs within air quality nonattainment and maintenance areas to perform air quality conformity tests (also called transportation conformity tests) prior to the adoption of RTPs and Transportation Improvement Programs (TIPs). The conformity requirement ensures that federal approval and funding go to transportation activities that are consistent with air quality goals.

A nonattainment area is one that the U.S. Environmental Protection Agency (EPA) has designated as not meeting certain national ambient air quality standards. A maintenance area is one that had a history of nonattainment, but is now meeting national standards. A conformity test is a demonstration that transportation plans, programs, and projects are consistent with the State Implementation Plan (SIP) for attaining the air quality standards. It is a determination that motor vehicle emissions will remain within emission budgets established within the SIP.

If an MPO fails to adopt a new RTP or TIP that stays within the motor vehicle emissions budgets in the SIP, the area faces what is known as a conformity lapse. During this period, the MPO cannot approve funding for new transportation projects or new phases of previously funded transportation projects except for those projects that are adopted as transportation control measures in the SIP or are otherwise exempt from conformity as air quality–neutral activities. Large urbanized areas stand to lose hundreds of millions of dollars in federal transportation funding if they cannot meet conformity requirements.

Conformity requirements have never resulted in loss of federal transportation funds, which has caused some observers to view them as toothless. MPOs have always been able to pass the conformity tests for whatever pollutants and years are at issue. This is mainly because technology is producing cleaner running vehicles with each passing model year. Atlanta famously suffered a conformity lapse that led to years of project delays, but funding was ultimately restored.

Conventional practice

The following determination from the Philadelphia area is typical of conformity in practice, even for an area that is not meeting NAAQS standards:

> Air quality in the DVRPC region does not meet the standards for two of these pollutants: ground level ozone and fine particulate matter (PM2.5). Therefore, the Clean Air Act requires DVRPC to demonstrate that the transportation projects contained in the TIPs and Plan do not make the region's air quality worse or impede the region's progress toward meeting the NAAQS. The process of this demonstration is referred to as transportation conformity.
>
> DVRPC demonstrates transportation conformity by using a travel demand model to estimate the motor vehicle emissions from all of the major regional projects in the TIPs and Plan and comparing those emissions against budgets or limits established by the states. This process is conducted in close coordination with an interagency consultation group, which is comprised of state and federal regulatory environmental, transportation, and transit agencies. DVRPC has successfully demonstrated the transportation conformity of *Connections 2040* and the Pennsylvania and New Jersey TIPs in accordance with the corresponding state implementation plans and Clean Air Act requirements.
>
> (Delaware Valley Regional Planning Council, 2013, p. 112)

Best practice

Other best practices look to historical precedents for guidance. This best practice charts new territory.

In April 2007, the U.S. Supreme Court ruled in *Massachusetts v. EPA* that the Clean Air Act gives EPA the authority to regulate emissions of greenhouse gases, if they are indeed a threat to human health and welfare. At the time, the court directed the agency to review the latest scientific evidence on climate change in order to make a determination.

In December 2009, EPA Administrator Lisa Jackson announced that the agency had finalized its finding that greenhouse gases, including carbon dioxide, pose a threat to human health and welfare. The ruling allows EPA to begin regulating GHG emissions from power plants, factories, and major industrial polluters. It also, in principle, allows EPA to begin regulating emissions from motor vehicles.

The book *Growing Cooler: The Evidence on Urban Development and Climate Change*, calls for such regulation under the air quality conformity rule.

> The obvious and best way for EPA to respond (to the Supreme Court ruling) is to extend transportation conformity requirements from criteria pollutants to GHGs [greenhouse gases]. Under such a system, state and local governments would be required to adopt mobile source GHG emission reduction budgets (like the emissions budgets for other pollutants) that demonstrate reasonable progress in limiting emissions. Currently, regions that fail to develop transportation plans consistent with "reasonable further progress" goals risk curbs on federal transportation funds. Withheld funds could be used to reward states and MPOs that effectively reduce per capita VMT.
>
> Although we acknowledge that, to date, land use and transportation demand management (TDM) policies generally have not played a significant role in meeting regional conformity requirements, we believe that comprehensive strategies aimed at GHG reductions would be more successful and less easily circumvented. Responsibility should be "nested" so that the federal government is responsible for the GHG impacts of federal transportation spending and state and local governments bear responsibility for the GHG impacts of their transportation spending.
>
> (Ewing et al., 2008, pp. 135–136)

This best practice requires that metropolitan regions be assigned CO2 budgets for motor vehicle emissions. EPA could establish such budgets based on the need to reduce overall emissions by something like 80 percent below 1990 levels by 2050. Logically, whereas transportation might not be required to do as much as other sectors to reduce overall emissions, it should be contributing something to the effort under a national plan. Given VMT and VMT per capita data, RTPs do not appear to be doing their fair share to stabilize climate in the long term (see Chapter 3). This is one conformity requirement that would not be easily met. It could shift funding away from projects that generate additional VMT, namely highway capacity projects.

Whereas this proposed best practice may seem radical, it is similar to what is being done in California under SB 375, the Sustainable Communities and Climate Protection Act. The California Air Resources Board issued greenhouse gas reduction targets to MPOs, targets they are required to meet through the development of Sustainable Communities Strategies as part of their RTPs. Oregon has an analogous law, SB 1059, the Oregon Sustainable

Transportation Initiative. As applied to Portland, SB 1059 actually has more teeth because it regulates local land use plans whereas SB 375 specifically exempts local land use plans from the requirement of consistency with the regionally adopted SCS.

Environmental justice

Background

Historically, low-income and minority residents faced disproportionately negative impacts from highway projects run through their communities. Further, these residents and communities were excluded from transportation policy setting or decision making and so did not receive a fair share of the benefits of transportation investments.

In response to these and other inequities, federal laws, rules, and regulations now promote environmental justice. Environmental justice is defined as the fair treatment of people of all races, cultures, and incomes with respect to the development, adoption, implementation, and enforcement of laws and policies.

The concept of environmental justice is rooted in Title VI of the Civil Rights Act of 1964 which prohibits discriminatory practices in programs and activities receiving federal funds. Title VI bars intentional discrimination, but also bars unjustified disparate impacts on EJ populations.

Executive Order 12898 – Federal Actions to Address Environmental Justice in Minority Populations and Low-Income Populations – was issued by President Clinton in 1994. Executive Order 12898 directs each Federal agency to develop an EJ strategy to implement its requirements. The federal DOT EJ strategy includes implementable action items that reflect DOT's commitment to EJ principles and integrating those principles into DOT programs, policies, and activities. DOT issued its original EJ strategy in 1995 and has updated it periodically since then.

Conventional practice

Whereas the federal metropolitan transportation planning rule makes reference to Title VI (23 CFR 450.334), it provides no guidance on how it is to be applied. The conventional practice of RTPs was to finesse environmental justice issues; many RTPs used to make no reference to them at all. It is now becoming more common, however, to see entire chapters of RTPs dedicated to environmental justice (Indianapolis Metropolitan Planning Organization, 2011, Chapter 11; New York Metropolitan Transportation Council, 2013, Appendix 4; Regional Transportation Commission of Washoe County, 2013, Chapter 9). Although some plans do not quantify environmental justice impacts, most make use of public outreach meetings and map EJ areas in regard to transportation projects.

> There are 3,082 census tracts within the NYMTC planning area of which 30 percent were identified as Communities of Concern. The total of 924 Communities of Concern represents census tracts with both 1) a minority population of 56 percent or greater, and 2) persons below the poverty level of 15 percent or greater. These census tracts were found in every county except Putnam County. New York City has the largest share of Communities of Concern (855 tracts), while the Lower Hudson Valley and Long Island regions have fewer Communities of Concern (47 and 22 tracts, respectively).
>
> (New York Metropolitan Transportation Council, 2013, Appendix 4–6)

Best practice

There are three fundamental EJ principles:

1 To avoid, minimize, or mitigate disproportionately high and adverse human health and environmental effects, including social and economic effects, on minority populations and low-income populations.
2 To ensure the full and fair participation by all potentially affected communities in the transportation decision-making process.
3 To prevent the denial of, reduction in, or significant delay in the receipt of benefits by minority and low-income populations.

The RTPs reviewed contain a number of applications of these principles. The most basic is the quantification of outcomes of plan investments on sub-areas with concentrations of the low-income and minority residents, and direct comparisons to sub-areas without such concentrations. In the development of its RTP, the Southern California Association of Governments utilized a number of performance measures designed to assess the overall equity of its investments:

> SCAG has established itself as a leader in environmental justice analyses and has been recognized for its technical approach to understand the benefits and burdens our regional plan. Each planning cycle presents new and emerging concerns for the region to address. For example, in the 2008 RTP, SCAG analyzed accessibility to public parks including the distribution of parks by income and park accessibility by travel mode and income. In keeping with the trend of developing robust environmental analyses, the current RTP/SCS analyzes impacts from rail transport, exposure to pollutants along heavily traveled corridors, gentrification and displacement, and impacts from revenue generating mechanisms such as a VMT fee.
>
> (Southern California Association of Governments, 2012, p. 182)

Table 6.5 Demographic Categories Used in Environmental Justice Analysis

Ethnic/Racial/Other Categories (persons)	*Income Categories (Households)*
White (Non-Hispanic)	Below Poverty Level
African-American	Income Quintile 1 (lowest)
American Indian	Income Quintile 2
Asian/Pacific Islander	Income Quintile 3
Hispanic (Latino)	Income Quintile 4
Other Racial Categories	Income Quintile 5
Disabled/Mobility Limited	
Age 65 and Above	
Non-English speaking	
Individuals without High School Diploma	
Households without a car	
Foreign-Born Population	
Young Children 5 and Under (Provided in Additional Analysis/Data)	
Sensitive Receptors: Hospitals, Daycare Facilities, Schools, Senior Centers, Parks/Open Space	

Source: Adapted from Southern California Association of Governments (2012, p. 187).

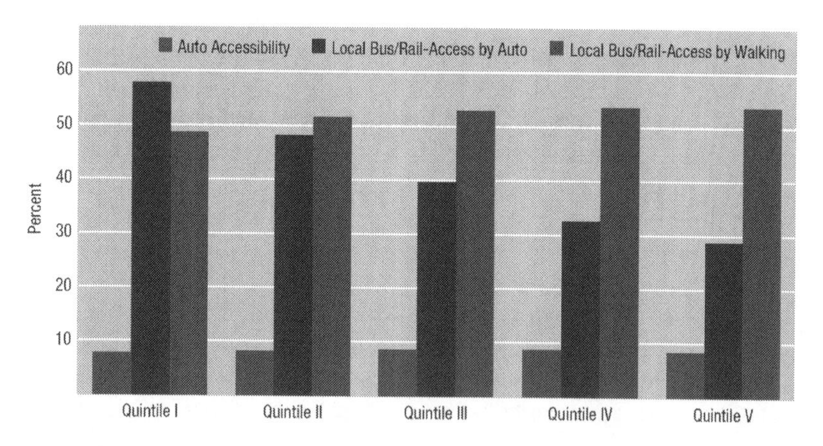

Figure 6.46 Comparison of Employment Accessibility Improvements by Travel Mode and Income Category (Plan vs. Baseline, 2035).

(Southern California Association of Governments, 2008, p. 177).

Other exemplary practices include setting aside funds specifically for mobility initiatives in disadvantaged communities, and making extraordinary efforts to engage EJ groups in the planning process. The San Francisco 2035 Plan committed an additional $400 million for transportation options in low-income communities. The MPO's Lifeline Transportation Program supports projects that address mobility and accessibility needs in low-income communities throughout the region. The Austin MPO, in coordination with the regional Environmental Justice Working Group, conducted two surveys designed to gather EJ community opinions related to transportation. The "Transportation Needs Survey for Environmental Justice Populations in the CAMPO Area" focused on concerns, safety issues, and solutions.

Transportation and land use modeling

Background

The long-range transportation planning process relies on forecasts of future travel patterns, which in turn rely on forecasts of future land use patterns because it is the need to go from one land use to another that generates travel. The conventional model structure is shown in Figure 6.47, taken from one of the RTPs. Future land use patterns are sometimes modeled mathematically but more often determined through a process of negotiation among participating governments based on their future land use plans. Future land use patterns are usually some version of trend development, which is usually some version of sprawl because the trend since World War II has been to sprawl.

These future land use forecasts are fed into a conventional four-step travel demand model, which first forecasts how many trips are produced in and attracted to small sub-areas called traffic analysis zones (trip generation), then forecasts flows among TAZs based on numbers of productions and attractions plus travel time between zone pairs (trip distribution), then splits flows among modes of travel based on demographic factors and relative

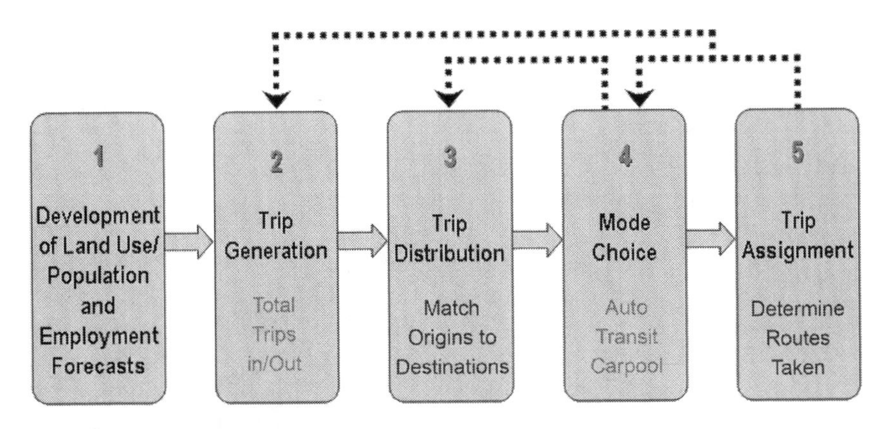

Figure 6.47 Conventional Travel Demand Forecasting Process.
(Regional Transportation Commission of Washoe County, 2008, p. 1–4).

travel times by alternative modes (mode choice), and then assigns flows to transportation networks such that aggregate travel times are minimized (network assignment).

Conventional practice

The conventional process has been roundly criticized primarily for failing to capture land use-transportation interactions in all their complexity. In trip generation, the trips forecasted are only those made by vehicle, when a significant number of trips are by walk or bike. Trips are treated as unlinked, when a majority of trips are actually part of multipurpose "tours." Vehicle ownership, a major input to trip generation, is often extrapolated without regard to D variables (density, diversity, design, destination accessibility, and distance to transit). In trip distribution, the number of trips forecasted to stay within zones disregards neighborhood land use patterns and street network configurations within zones, and measures of accessibility used to distribute trips among zones assume trips are unlinked. In mode choice, walk and bike trips are often not forecasted at all, and among motorized trips by auto or transit, mode splits are treated as independent of D variables within zones of origin and destination. To convert from daily traffic volumes to peak hour or peak period volumes to be assigned to the network, the conventional model uses fixed factors rather than accounting for "peak spreading" as sub-areas become more urbanized.

Best practice

Given the importance of land use and travel forecasting in the long-range transportation planning process, surprisingly little is said in RTPs about the analytical tools used to make these forecasts. This discussion therefore extrapolates from what is stated, in light of the literature generally.

You might assume that the best practice would involve modeling future land use patterns objectively rather than relying on the subjective and political process of negotiation among local jurisdictions. This isn't necessarily the case. Two of the featured MPOs (Dallas-Ft. Worth and Kansas City) once used the land use forecasting tool, DRAM/EMPAL. This forecasting process was described this way in the Kansas City RTP:

> DRAM and EMPAL were both calibrated over the 1980 to 1990 period. They were then run successively, with DRAM allocating households based on employment location then EMPAL allocating employment based on household location – to step forward in time to the end of the forecast period, which was originally 2020. The resulting trend forecast was reviewed by MARC's Technical Forecast Committee for consistency with local plans and adjusted as necessary.
>
> (Mid-America Regional Council, 2005, p. 2–2)

The problem with this approach, which caused the Kansas City MPO to abandon it, is the following:

> While good for forecasting historical trends into the future, the structure of DRAM and EMPAL was not appropriate for assessing how trends might change in response to policy changes like the imposition of impact fees on new development or significant new investments in the region's urban core. Therefore, the Technical Forecast Committee modified the trend forecast to represent its judgment concerning the likely impact of existing policy and investment initiatives. It was this policy-oriented forecast that was adopted by the MARC Board of Directors for use in transportation planning.
>
> (Mid-America Regional Council, 2005, p. 2–2)

There appear to be two good alternatives to negotiated forecasts and trend-based model forecasts. One is to use newer policy-sensitive, behaviorally oriented land use forecasting models like UrbanSim or PECAS. This is the approach taken in Salt Lake City with UrbanSim (now referred to as the Real Estate Market Model) and in Baltimore with PECAS:

> UrbanSim is a state-of-the-art approach to forecasting future land-use growth with growth forecasts influenced by the quality of the proposed transportation system. By coupling UrbanSim with the regional travel demand model system, a range of land use and transportation policy interventions are combined into policy 'scenarios', and the systematic effects of these intervention strategies can be expressed in terms of projected urban development outcomes and the quality of the transportation system.
>
> (Wasatch Front Regional Council, 2007, p. 39)

Likewise, from the 2007 RTP for Baltimore:

> Since the early 1990s, the BRTB has sought to fully integrate a land use model with the Baltimore Regional Travel Demand Model. In order to develop such a model, the BRTB has tested numerous software packages and developed databases needed to implement an econometric land use model. The BRTB has chosen to work with the Production Exchange and Consumption Allocation System, or PECAS model, to investigate the link between transportation infrastructure development and the price and movement of commodities and floor-space in the region.
>
> (Baltimore Regional Transportation Board, 2007, p. 23)

Unfortunately, due to fiscal issues, BRTB did not update PECAS for its latest RTP forecasts. It hopes to resume its PECAS forecasts should funding become more certain.

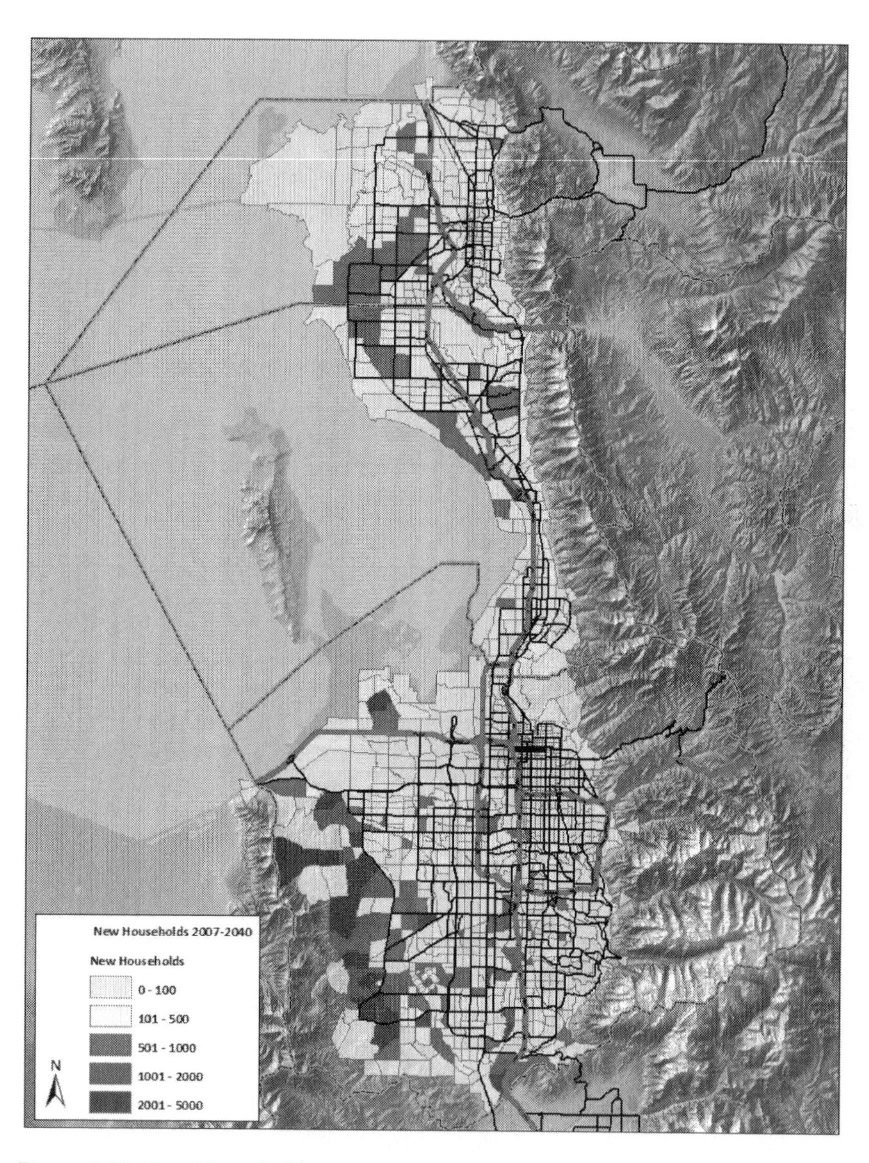

Figure 6.48 New Households 2007–2040.
(Wasatch Front Regional Council, 2011, Appendix A, p. 13).

The other approach is to arrive at a normative future land use forecast through a scenario planning process (see Scenario Planning). This is the approach taken in the Sacramento RTP, using the planning support tool I-PLACES3S. The Sacramento MPO describes the approach this way:

> The problem with scenario planning is that it's technical; without a software tool like I-PLACE3S, scenario planning can be complex and intimidating, and can devolve either into planning by guesswork (with little or no technical information because it is simply too costly and impractical to create), or the reverse, a simple accounting of numeric data. Like any tool, I-PLACE3S must be used properly to achieve the desired results.
>
> (Sacramento Area Council of Governments, 2010, p. 1)

The I-PLACE3S tool was instrumental in the Blueprint Project from 2002 to 2004 and in the development of the Metropolitan Transportation Plan 2035 land use allocations. This tool is no longer in use while other scenario tools are being developed that will continue and extend the ability to create detailed land use scenarios and assess their impacts. The capability continues to be important in the Blueprint implementation efforts at SACOG and several member and partner agencies. SACOG staff and member agencies also use it to develop land use scenarios that feed into travel and air quality modeling (Sacramento Area Council of Governments, undated).

The other area of best practice is in travel demand modeling. The Salt Lake City travel demand model has "several advanced features that place it on the cutting edge of the improved modeling methodology needed to meet the requirements of SAFETEA-LU and the Clean Air Act Amendments of 1990" (Wasatch Front Regional Council, 2011, Appendix C, p. 1). We are not told what they are, though. The St. Louis model has been given a "major overhaul," but we are not told what that entailed (East-West Gateway Council of Governments, 2007, p. 4). The Burlington model "incorporates several advanced features including the ability to estimate bus, commuter rail, walk/bike and shared and single occupancy vehicle trips, and sensitivity to the effect transportation projects have on where trips are made" (Chittenden County Metropolitan Planning Organization, 2013, p. 159). Again, we do not know what these features are. We suspect in all three cases the travel demand models have remained quite conventional.

Whereas all of the above are doubtless useful modeling innovations, what really distinguishes conventional practice from best practice is the ability to account for the effect of the D variables (density, diversity, design, destination accessibility, and distance to transit) on vehicle ownership, trip rates, trip distances, mode choices, and peak-to-daily ratios. In this regard, the best practice may be accomplished either by post-processing outputs of more conventional models, or by developing state-of-the-art micro-simulation models. The Wasatch Front Regional Council (WFRC) in Salt Lake City is in the process of developing models that better account for car shedding, walk and bike trips, intrazonal travel, and peak spreading.

The models currently used in most regions are underspecified, which is to say that important variables are omitted. In particular, conventional models fail to fully account for local land use patterns, street network designs, and urban design features – indeed, the entire built environment at the scale of a neighborhood or activity center. In many four-step models, vehicle ownership is treated as a function of sociodemographic variables only (or largely), and the phenomenon of car shedding as the built environment becomes more compact is not accounted for. In many models, only trips by vehicle are modeled, and trip rates are related only to sociodemographic characteristics of people, not characteristics of place. Bicycling, in particular, is seldom treated as a separate transportation mode. In nearly all four-step models, households, jobs, and other trip generators are assumed to be located at a single point, the zone centroid, rather than spread across the traffic analysis zone, and the entire local street network is reduced to one or more centroid connectors to the regional street network. This precludes the modeling of intrazonal travel in terms of the local built environment. In the conventional model, daily traffic volumes are factored to obtain peak hour volumes without regard to the phenomenon of peak spreading as development becomes more concentrated and congestion increases. Whereas there are other ways in which conventional travel demand models fail to account for land use-travel interactions, these four are the focus of the WFRC model enhancements.

Models will be calibrated with data from a 30-region household travel database, the largest household travel database of its sort ever assembled. This database has been linked to built environmental data for traffic analysis zones in the 30 regions. The models will be validated using k-fold cross validation, which provides more accurate measures of uncertainty in the model than a leave-one-out approach.

An emerging development in transportation planning is activity-based modeling. Similar in some respects to the traditional four-step travel demand model – both generate activities, destinations for activities, travel modes, and a specific network of facilities for each trip – activity-based models build on the existing standard by incorporating constraints of space and time and the linkages among activities and travel. These elements are considered for individuals as well as multiple people in a household (Castiglione, Bradley, & Gliebe, 2015).

Activity-based models more closely represent the decisions made by travelers and should, therefore, provide better forecasts of expected travel patterns. A second advantage of activity-based models is that they can be used to evaluate policies that are difficult to test in trip-based models. For example, Castiglione et al. (2015) tout the robust capabilities of activity-based models in evaluating pricing scenarios. Because they function at the person level and account for how individuals travel throughout the day, activity-based models are more sensitive to pricing policies that vary by time of day. A final advantage of activity-based models is that they produce more detailed performance metrics.

Implementation cost is a significant factor limiting the use of activity-based models in many MPOs, but in recent years costs have declined to the point that activity-based models are now competitive with traditional trip-based models. The main limitation now with activity-based models is limited expertise in most MPOs. Activity-based models are data-intensive and the skills required to build, test, and implement them are not well dispersed.

This may explain why activity-based models are not frequently used by MPOs, even though they have been discussed in academic literature since 1978. Cambridge Systematics, Inc. (2015) conducted an analysis of the 23 MPOs with populations of 1.5 million or greater and found that only six (New York, San Francisco, San Diego, Denver, Sacramento, and Columbus) were actively using activity-based models. No similar analysis has been conducted for MPOs with smaller populations, but we suspect that activity-based models are even less frequently used in these MPOs. The tide could finally be shifting for activity-based models, however. Of the 23 MPOs reviewed by Cambridge, all but six are studying or developing activity-based models for use in future planning efforts.

Beyond the analytical tools that are used, another best practice involves the assumptions that are made about the distant future. All plans look forward at least 20 years, and many look forward 30 years or more (most have horizon years of 2040 or later, one has 2060). Odds are that the devastating impacts of climate change will have forced world governments to take strong action to curtail greenhouse gas emissions. Also, major demographic shifts will have occurred (see Demographics). Some future scenarios should relate to the very different world our grandchildren will inherit. Finally, the cost of transportation fuels, for those who still drive fossil-fueled cars, will likely have risen dramatically (see Goals). One RTP that at least nods to the possibility is Los Angeles's:

> Travel demand forecasts generally assume that the future will include an abundant and relatively inexpensive supply of transportation fuels. However, this assumption is in question based on the International Energy Agency's (IEA) 2010 World Energy Outlook.
>
> The IEA forecasts that the emerging economies of India and China will drive global energy demand higher. The IEA further states that China overtook the United States in 2009 as the world's largest energy consumer and their consumption will continue to grow. If governments act more vigorously to increase fuel efficiency and promote demand for alternative fuels, the demand for oil will decrease, avoiding price increases and supply disruptions.
>
> However, if fuel prices continue to increase, it would have a ripple effect on numerous areas including construction costs, gas tax revenue, travel and aviation demand, air emissions, mode choice and growth

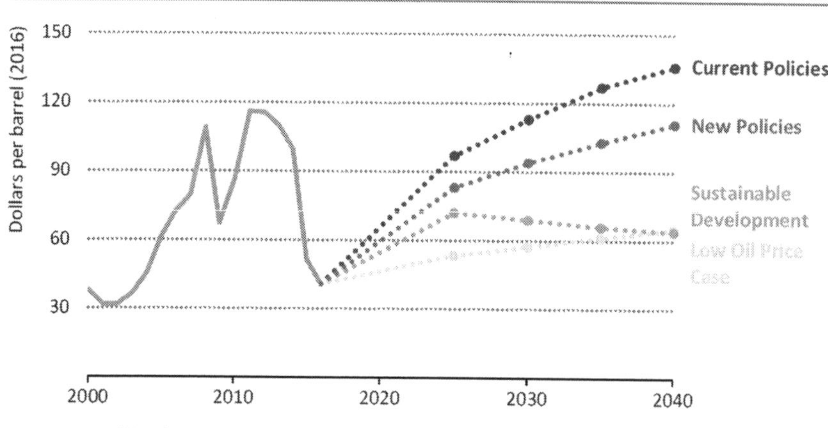

Figure 6.49 Average IEA Crude Oil Import Price by Scenario and Case. (International Energy Agency, 2017, p. 54).

> patterns. In response, the 2012 RTP/SCS supports the increased adoption of near zero and zero emission technologies to lessen the region's exposure to fossil fuel price spikes resulting from an uncertain energy future and reduce GHGs and emissions of criteria pollutants.
>
> (Southern California Association of Governments, 2012, pp. 29–30)

Public engagement

Background

The traditional Western democratic process is characterized by the general public contributing to their governance through the election of policy makers who best represent their views. This is done at a variety of jurisdictional levels, depending on the type of democracy (i.e., federalism, centralism, etc.). Representatives are subsequently trusted to make policy decisions with their constituents' views in mind, but without direct consultation with the constituents.

Public engagement is different in that the public is asked to participate in more of the minutia of decision making. This process requires a greater degree of involvement, and those participating are expected to be more aware of the issues related to the decisions to which they are contributing. There have been significant changes to the process of public engagement in the United States throughout the past few decades.

The roots of public participation can be traced to 1964 with the Economic Opportunity Act, which led to the creation of community action agencies (CAAs) and the community action program (CAP). The explicit goal of the CAP was to achieve the "maximum feasible participation" of community members. However, CAAs soon took on a more economic focus, as opposed to their original orientation towards political organization. The diversion from the CAP's original purpose was made more concrete with the passage of the Federal Community Self-Determination Act of 1969. It is speculated that this and later legislation was a response to municipal governments' fear of the increasing power accumulated by community organizations and their ability to express an unprecedented level of autonomy and self-governance (DeFilippos, 2008).

The disheartening direction of policy during this period, however, was responded to with a significant intellectual and grassroots movement for increased participation at the neighborhood level (Arnstein, 1969). Writers like Jacobs (1961) and Kotler (2005) argue that large governments are too complex in their institutional structure, and their broad over-arching policies do not appropriately reflect the needs of people at the community level. A number of organizing groups and advocacy coalitions arose during the 1970s that were relatively nonideological, but rather worked individually towards the general goal of local control, autonomy, and greater participation in the local decision-making process.

Although similar, it is important to distinguish between community organization and public engagement. Whereas they come from different root necessities, they achieve their goals in similar fashion. Both require active participation from individuals that are typically not a part of the traditional power structure or decision-making process. Community organization came from a general sentiment of dissatisfaction and the need for collective action to enact radical change in favor of those outside the traditionally entrenched authority. This is different from the more modern notion of public engagement. It has been stated that "participation has [recently] become a central feature of making and implementing policy" (Lane, 2005). The theoretical purpose, then, of the contemporary public engagement process is to improve decision makers' ability to plan and create policy that works for their constituents. It is assumed that the opinions and knowledge that is brought forth by participants provides a perspective that is novel and beneficial to policy makers.

Where does it leave public engagement today? *Bowling Together*, the cleverly titled piece by Coleman and Gotze (2001), uses Putnam's *Bowling Alone* view on the disappearance of social capital as a contrast to the ever-expanding avenues, made possible by the information age, through which individuals can participate in the planning and decision-making process. Social media and the proliferation of cheap communication have made it possible for those in power to broadcast the issues being discussed, and it has also opened up a seemingly infinite number of channels through which concerned citizens can voice their opinions on such matters.

Conventional practice

The Federal Highway Administration describes the conventional practice of public engagement as an example of what not to do. "Public participation is more than just a hearing, or one meeting near the end of the project development process." Per federal requirements, most MPOs now prepare elaborate public participation plans. They often fill an entire chapter or appendix of the RTP. It is hard to find examples of large MPOs that do only the bare minimum required by federal regulations. One MPO that seemed to do this in previous versions of the RTP, based on its own public process description, is the Indianapolis MPO:

> Key components of the public involvement program are the MPO newsletters and the MPO web site. The MPO staff presented the 2030 RTP 2009 Update process to the IRTC, as well as the subsequent draft list of cost-feasible projects. The 2030 RTP 2009 Update was also made available to the public during a public comment period where the Update was available at numerous public locations including libraries . . . The public review period as well as all public hearings is advertised on the MPO's website, in local newspapers, and at local libraries.
>
> (Indianapolis Metropolitan Planning Organization, 2009, p. 11)

Best practice

It is clear from the review of RTPs that some MPOs take public participation more seriously than others. The Boise RTP's public involvement objective is to "seek representation from the wider community, will reach an underserved population, will offer a range of educational opportunities, and provide public input to planners and decision-makers in a timely manner" (Community Planning Association of Southwest Idaho, 2010, p. 2–7). Generalizing across RTPs that are more engaging, the elements of best practice include:

Early and continuous

The Boise process came in four phases, from initial scoping with the Regional Transportation Task Force to final plan review and approval. Ultimately, 2,000 people were involved in the multiyear public involvement process, and they participated in workshops, cafés, and presentations during the first three phases of the process. The fourth phase was particularly creative. Rather than presenting the RTP in a traditional open house or public hearing setting, residents hosted dozens of meetings with their friends, peers, and/or colleagues to review and discuss the plan (Community Planning Association of Southwest Idaho, 2010, p. 2–9).

Intensive and in-depth

Rather than one-time affairs like public meetings or open houses, committees can be charged with ongoing responsibility for different aspects of the RTP. The Sacramento MPO held 16 focus groups with follow-up sessions that included a diverse group of stakeholders representing different constituencies and interests in regional growth, including agriculture, climate action planning, education, and seniors and aging populations, among others (Sacramento Area Council of Governments, 2012, p. 17).

Rich with relevant information

Participants in the Sacramento process were provided maps of nine transportation corridors and performance information from computer modeling associated with specific transportation investment packages. The performance indicators included vehicle miles traveled in congestion, vehicle hours of travel per household, and percentage of trips driving alone, carpooling, walking, bicycling, and using transit by 2035 (Sacramento Area Council of Governments, 2008, p. 86).

Varied in format

The New York process used all of the following: open houses; public workshops and planning sessions; transcription and translation services so those who wished to provide feedback were able to do so regardless of their ability to read, write, or understand English; an interactive website; and social media (New York Metropolitan Transportation Council, 2013, p. 7–3).

High tech and high touch

To engage a generation that relies on and is most comfortable with the Internet for information, a high-tech approach was developed in Austin that included a website with the ability to send e-blasts about community meetings; online surveys to identify overall priorities, preferences, and prioritization of projects; use of social media outlets that included Twitter, Facebook, and YouTube. For those without access to or an aptitude with the Internet, high-touch techniques encouraged interested parties to become engaged in the process through community workshops that introduced the planning process and solicited feedback with interactive mapping exercises; targeted outreach, specifically to environmental justice populations; display booths staffed by public involvement team members and MPO staff members to answer questions in a one-on-one format (Capital Area Metropolitan Planning Organization of Austin, 2010, p. 10).

NEWS RELEASE

For More Information:
Contact: Jon Coleman,
Executive Director

Phone # (517) 393-0342
Fax #: (517) 393-4424

TRI-COUNTY REGIONAL PLANNING COMMISSION
913 W. Holmes Rd., Suite 201, Lansing, Michigan 48910 • Area Code 517/393-0342

FOR IMMEDIATE RELEASE

COMMISSION ANNOUNCES TOWN HALL FORUMS ON REGIONAL 2035 TRANSPORTATION PLAN GOALS AND OPTIONS.

If you had the opportunity to help local officials determine how to best invest over $2.8 billion dollars on the region's transportation system between now and 2035, what goals and priorities should be established for use of those funds?

That's the essence of an important series of public town hall forums slated for December 3-9 in four locations around Clinton, Eaton and Ingham counties as the Tri-County Regional Planning Commission begins to update the region's transportation plan to 2035.

Figure 6.50 Public Outreach.
(Tri-County Regional Planning Commission, 2010, p. 5–11).

Note

1 C.P. Flynn and L.J. Glazer, "Ten Cities' Strategies for Transportation Demand Management," *Transportation Research Record 1212*, 1989, pp. 11–23; KPMG Peat Marwick, *Status of Traffic Mitigation Ordinances*, Urban Mass Transportation Administration, Washington, DC, 1989; E. Ferguson, "Transportation Demand Management: Planning, Development, and Implementation," *Journal of the American Planning Association*, *56*, 1990, pp. 442–456; T.H. Higgins, "Guidelines for Developing Local Demand Management or Trip Reduction Policies," *Transportation Research Record 1280*, 1990, pp. 11–21; Center for Urban Transportation Research (CUTR), *Commute Alternatives Systems Handbook*, University of South Florida, Tampa, 1992, pp. 57–65; E. Sanford and E. Ferguson, "Overview of Trip Reduction Ordinances in the United States: The Vote Is Still Out on Their Effectiveness," *Transportation Research Record 1321*, 1991; E. Ferguson, C. Ross, and M. Meyer, "Transportation Management Associations: Organization, Implementation, and Evaluation," *Transportation Research Record 1346*, 1992, pp. 36–43; and R. Ewing, "TDM, Growth Management, and the Other Four Out of Five Trips," *Transportation Quarterly*, *47*, 1993a, pp. 343–366.

7 The end of mobility

To this point, we have discussed the history and present practice of regional transportation planning in the United States. In Chapter 1, we considered the shift from the largely self-contained American community of yesterday to the megalopolitan regions of today using the Salt Lake City area for context. Although people now regularly travel outside of the communities where they live, most planning in the United States still occurs at the municipal level, often frustrating regional planning efforts that attempt to be more congruent with the daily activities and travel patterns of the average American.

In Chapter 2, we explored the historical purpose of cities. Cities facilitate the exchange of goods and services. Transportation planning has curiously

Photo: Keith Bartholomew.

emphasized the trips between exchanges rather than the exchanges themselves. In focusing almost exclusively on mobility, regional governments ignored concepts like accessibility and livability that are arguably more important to people. Disabled people, the young, and the elderly are particularly poorly served by mobility-related policies at the regional level.

The shift toward accessibility occurred in the early 1990s with the passage of ISTEA, the Intermodal Surface Transportation Efficiency Act of 1991. For the first time, the federal government called attention to the connection between land use and transportation planning and prioritized projects that limited the effects of induced demand. Subsequent legislation in the form of TEA-21, SAFETEA, MAP-21, and FAST Act has not been as effective, but metropolitan planning organizations (MPOs) throughout the United States have to a greater or lesser extent assumed the mantle of accessibility- and livability-based transportation planning, with measures like vehicle miles traveled (VMT) outpacing traditional mobility-based measures like level of service (LOS) and delay.

In Chapter 3, we conducted a quantitative analysis of the regional transportation plans (RTPs) of 110 MPOs. This analysis included population and employment growth, daily VMT and VMT per capita, delay and delay per capita, mode shares, and expenditures on highways, alternative modes, and transportation system management. Notably, a quarter of all development in the average MPO will be built between now and the horizon year listed in most RTPs, generally 20 to 40 years away. This means we have an opportunity: to continue doing sprawling, autocentric development, or to shift toward compact, multimodal development that better serves users of a variety of transportation modes.

In Chapter 4, we considered an emerging trend in regional transportation planning: scenario planning. Scenario analysis borrows from business and military strategic planning and improves on the alternatives analyses that were popular with MPOs in the 1970s. More than just forecasting how a region is likely to function in the future based on current trends, scenario planning allows regions to consider a variety of alternative possible futures. Armed with a knowledge of possibilities, regions are then empowered to adopt goals and strategies most likely to realize the preferred scenario.

In Chapter 5, we looked at the historical growth of VMT in virtually every region in the United States. Highway capacity, fuel prices, transit service, and income were shown to affect VMT. Most influential is destination accessibility, one of a suite of "D variables" (also including density, diversity, design, and distance to transit) important in improving the overall accessibility of a site, neighborhood, city, or region. The integrated approach used here – new to this chapter – provided a platform for understanding how different policy options might work in practice.

Finally, in Chapter 6, we conducted a qualitative analysis of regional transportation plans. Conventional practices in many regions were juxtaposed with best practices in some exemplary regions, including San Francisco and

Portland. In exemplary regions, mission statements, goals, objectives, and performance measures are clearly defined, ambitious, and progressive. Scenario planning is the norm, and the connection between land use and transportation is well understood and considered holistically. Exemplary regions channel growth into employment and population centers, advocate smart growth, and consider demographic shifts an opportunity to adjust current practices rather than maintain policies that won't work.

Exemplary regions focus on street connectivity and avoid endlessly expanding highway capacity. Maintenance of existing infrastructure is prioritized over new construction, and transportation system management and travel demand management are fully funded. Improvements to public transit and bicycle and pedestrian facilities are seen as more than a mere afterthought. Innovative approaches to securing funding, selecting projects, and modeling land use and transportation are commonly seen in exemplary regions. Finally, these regions are more likely to prioritize air quality conformity, environmental justice, and effective public engagement.

As outlined in this book, transportation professionals have been in somewhat of an existential crisis for a couple of decades. Traditionally tasked with keeping vehicles moving fast, they have now learned that not only is that an impossible objective, but that it is not what their constituents really want. Although people may talk the talk of "mobility" and how to increase it, what they really want is some safe, healthy, pleasant place to live and ready access to opportunities. Providing that is a much more complicated venture that just maintaining a specified level of service on the highway network (as difficult as that it is).

State departments of transportation and metropolitan planning organizations (MPOs) dole out billions of dollars annually for specific roadway construction projects to widen existing highways or build new corridors. Although billions have been spent on added capacity throughout the past few decades, each region in the country has experienced increased congestion over this period. For all but eight of the 101 urbanized areas in the Texas Transportation Institute's Urban Mobility Scorecard (Schrank et al., 2015), annual delay per commuter more than doubled between 1982 (the first year in the series) and 2014 (the last year in the series). For all but one urbanized area, annual delay per commuter increased by more than 40 percent over this same period.

In this closing chapter, we look forward to the future to see how transportation professionals will finesse this mission shift toward an agenda that is more about livability and opportunity and less about motion. Key to this transition is how transportation is measured. Peter Drucker, the famous business writer and educator, turned the old saw "what gets measured gets done" into a slightly more apt version for this discussion: "What gets measured gets managed." Given that the customary way to assess the workings of highways, roads, or transit systems has been to meter the speed of vehicles, it is not terribly surprising that speed, and its counterpoise congestion, became transportation's obsession. If transportation is to transition from mobility to livability and accessibility it will need to start measuring something else, something more focused on how well peoples' life needs are being met.

The traditional way of measuring roadway capacity for systems planning and traffic operations is through levels of service or LOS. LOS measures roadway performance on an A to F scale, like a student's report card. LOS A represents free flow, LOS F gridlock. Like grade-preoccupied students, many transportation planners and traffic engineers pursue good "grades" with focused determination. But if the recent trends in transportation consumption are any indication, people are not as concerned about speed and capacity as maybe they once were.

Despite 60-plus years of constant growth in travel, frequently at rates several times the rate of population growth, the early 21st century has seen a marked decline in per person VMT. True, the economic downturn of the Great Recession and higher gas prices in 2008 contributed to this result, but the phenomenon now referred to as "peak car" began before those milestones and has continued since then. The first sign of peak car in the United States was in 1992 when VMT per capita rates in Washington State first plateaued and then began to fall (Garceau, Atkinson-Palombo, & Garrick, 2015). By 2011, per person VMT was 13 percent lower than in 1992, this despite a 24.6 percent increase in GDP per person during the same period.

Within transportation planning circles, this was mind-blowing. For nearly a century, economic activity and vehicle travel had been linked: when GDP increased, so did VMT with the increased travel reflecting more economic activity (McMullen & Eckstein, 2012). The relationship was so tightly bound that, for transportation planners, economic indicators became the primary predictor of travel demand. What was signaled by Washington's VMT data was nothing short of revolutionary – a possible decoupling of economic conditions and transportation consumption. How could this occur? Upon reflection, it makes sense that at some point these two spheres would diverge. As discussed in Chapter 2, transportation is, fundamentally, a derived demand – a transaction cost required to engage in economic or social interaction. As greater efficiency rolls through the rest of the economy, it seems logical that at some point, the influences rewarding greater efficiency would work to reduce transportation, just as they have minimized other transaction costs.

But what about the rest of the world? Whereas Washington State's news is interesting, Washingtonians have, at times, been viewed as being somewhat out of the norm compared to the rest of the country. Joel Garreau (1981), for example, once referred to the state as "Ecotopia," a term the *Oxford English Dictionary* defines as "an ecologically ideal region or form of society, generally viewed as imaginary." It turns out that as falls Washington State, so falls the rest of the United States. A state-by-state analysis shows that by 1999, six additional states had joined Washington. In 2004 – four years before the recession – half of the states were showing signs of having peaked. By 2011, only Mississippi and North Dakota were not showing peak conditions. In that same year, 40 states posted lower VMT per capita levels than the previous year (Garceau, Atkinson-Palombo, & Garrick, 2015). Moreover, similar trends were being observed in many other Western industrialized nations.

The idea of market saturation of car ownership, licensing, and use started to appear in the academic literature in the 1960s and 1970s (see Goodwin & Van Dender 2013 for citations). With an impressive degree of prescience, a plurality of the projections targeted the first decade of the 21st century as the likely date that peak car would occur. At the center of these projections was the observation that average commute travel times had been impressively stable, leading to some to refer to a "commuting constant." Looking at historical data, some researchers observed this constant in Western European cultures as far back as Elizabethan England (SACTRA, 1994). At the heart of this notion is basic human behavior and the limits of the diurnal clock: given a 24-hour day and human biological and physical needs, people have only so much time to spend on commuting. From commuting, researchers extended the idea to other travel purposes, creating the concept of a "travel time budget." As travel technologies and infrastructures increased travel speeds, travel distances tended to increase (constant travel time x faster speed = increased distance), especially as individual and societal affluence increased. But at a certain point, the limits of technology and economy were bound to lead to a plateauing of travel distances. That point was reached in the early 2000s, right on schedule. Between 2000 and 2010, the degree of urban sprawl in U.S. urbanized areas decreased in as many areas as it increased, this after decades of secular increases (Hamidi & Ewing, 2014).

In addition to observing the apparent break between economic growth and travel, the burgeoning peak car literature has coalesced around two major influences: the behaviors and preferences of teens and young adults and the growth of denser, mixed-use urban centers and alternative modes for traversing them. On the first of these issues, much attention has been focused on the seeming indifference of the members of the Millennial Generation (i.e., those born between 1983 and 2001) toward automotive travel, especially when compared to preceding 20th-century generations. Baby boomers, like us, were chomping at the bit to get licensed and on the road. For a generation that was largely reared in the safe and sedate suburbs, cars represented access to culture and social exchanges beyond what was available on the cul-de-sac. The first step toward those opportunities was getting a driver's license, and in the decades following World War II, approximately 85 percent of us had done that by the time we reached 18 (Delbosc & Currie, 2013). Millennials, by comparison, do not seem as hell bent to spring free from the nest, at least in a car. In a recently published study of car licensing data, Delbosc (2016) shows that in 1997 only 74 percent of 18-year-olds were licensed and there has been a steady decline in the percentage every year since. By 2011 (the last year data were available), the percentage had dropped to 60 percent – below two-thirds. Also impressive was the decline in 16-year-old licensing rates, going from 41 percent in 1995 to 31 percent in 2009.

It's not just Millennials who are changing their habits. Americans of all demographic stripes are shifting their real estate preferences toward what Christopher Leinberger (2009) calls walkable urbanism:

> places characterized by much higher density and a mix of diverse real estate types, connected to surrounding areas via multiple transportation options, such as bus and rail, bike routes, and motor vehicles. For those living or visiting a walkable urban place, everyday destinations, such as home, work, school, stores, and restaurants, are within walking distance.
>
> (Leinberger & Lynch, 2014, 6)

In 2011, we surveyed the "hedonic price studies" of transit- and pedestrian-oriented development (Bartholomew & Ewing, 2011), real estate types largely meeting Leinberger's definition of walkable urbanism. Hedonic analysis looks at the prices people are willing to pay for certain attributes contained within a consumer product. In real estate contexts, the technique teases out the price premiums (or penalties) associated with such features as extra bathrooms, large yards, and proximity to good schools. In our review, we focused on what people were willing to pay for proximity to transit, mixed uses, and pedestrian design. Although there was a great deal of variation study to study, virtually all of the analyses indicated that people were willing to pay more for these features. In some cases, the price premium was as high as 28 percent (Atkinson-Palombo, 2010). Since 2011, more than 80 additional hedonic analyses have been conducted showing similar outcomes. The phenomenon has become so predictable that in the Netherlands, researchers are starting to use hedonic-based models for government-required cost-benefit analyses of proposed new transit projects (de Graaf, Ghebreegziabiher, & Rietveld, 2012).

Hedonic studies uncover "revealed" preferences. "Stated" preferences tell the same story. There have been a number of attempts to assess citizens' attitudes toward alternative development via surveys. Perhaps the best national gauge of this demand is the National Survey on Communities, conducted for Smart Growth America (a nonprofit advocacy group) and the National Association of Realtors (NAR) (Belden, Russonello, & Stewart, 2004). In this survey, respondents were given a choice between suburban communities labeled "A" and smart growth communities labeled "B." Overall, 55 percent of Americans expressed a preference for community B. This community appealed to 61 percent of those who were thinking of buying a house within the next three years. Commuting time had a significant influence on respondents' preferences. About a third of the respondents said they would choose the smart growth design if commutes were comparable, whereas another quarter preferred such a design if it also meant being closer to work.

This survey was updated by the NAR in 2011 under the title "The 2011 Community Preference Survey: What Americans Are Looking for When

Deciding Where to Live." Over 2,000 Americans were surveyed. The survey covered policy issues, community needs, preferred community type, preferred housing type, and tradeoffs between house and lot size, walkability, and commute time. The survey concluded:

> most Americans would like to live in walkable communities where shops, restaurants, and local businesses are within an easy stroll from their homes and their jobs are a short commute away; as long as those communities can also provide privacy from neighbors and detached, single-family homes. If this ideal is not possible, most prioritize shorter commutes and single-family homes above other considerations.

In fact, given the same choice between communities A and B, 56 percent of respondents selected the smart growth community independent of commute times. This indicates a substantial shift in residential preferences between 2004 and 2011.

Fundamentally, all of these studies are looking at the same thing: the value that people place on ease of access to the activities and opportunities that provide value in life. In other words, accessibility. The fact that people are paying more for accessibility in their real estate investment choices tells us that this is what people really want, not mobility. That being the case, shouldn't transportation planners use measures that focus on the ease of access to desired interactions – accessibility – rather than just the ease of purposeless motion – mobility? What would those measures be?

Prognostication about accessibility measures has become something of a niche obsession among academics, with literature going back as far as 1931 (Pooler, 1995). In the ensuing decades, researchers have posited a number of measures, ranging from GIS-based cumulative opportunity indices (counting exchange opportunities within a defined geographic parameter), to gravity-type models (pitting the importance of exchange opportunities against the amount of travel time required to reach them), to utility measures (which quantify the benefit individuals get from destinations), to complex space-time prisms (three-dimensional constructs measuring individuals' range of possible geographic movement within specified time constraints).

In sorting through these competing systems, Geurs and van Wee (2004) have offered a five-dimension matrix for evaluating the suitability of accessibility measures for deployment in on-the-ground planning applications. First, an accessibility measurement tool needs to meet the theoretical requirements of accessibility. In other words, it should be able to capture effects of changes in transportation capacity and land use proximity – accessibility can be increased either by moving people to opportunities or by bringing opportunities closer to people. Second, it should be capable of being operationalized given the reasonable availability of data and resources. Third, it should be capable of reasonable interpretation, communication, and understanding by users and policy makers. Fourth, it should capture economic effects, both at the macro and micro levels. Finally, it should facilitate understanding of social impacts from policy decisions.

Whereas the accessibility measurement approaches noted above all have appreciable strengths under this matrix, they also suffer from significant limitations. Cumulative opportunity measures, for instance, do not distinguish between destinations' quality or desirability. Gravity measures produce numeric scores for individual destinations that are not comparable across different settings, making them less intuitive to understand (Levine, Grengs, & Shen, 2009). Utility measures are problematic because of the difficulty in gathering the necessary data and the range of socioeconomic differences between individuals leading to different travel values (Niedzielski & Boschmann, 2014). Space-time measures, while having a strong theoretical basis, involve data and analytical complexities that make them difficult to implement and interpret (Geurs & van Wee, 2004). None of these measures factor in the increasing role that information communication technologies (ICT) play in providing access to employment, education, culture, and commerce (van Wee, Chorus, & Geurs, 2012).

These shortcomings demonstrate why it has been difficult for MPOs to adopt accessibility as a functional framework for planning practice (Curl, Nelson, & Anable, 2011). With limited funding and staffing, many MPOs report limited interest or ability in deploying additional metrics aimed at accessibility (Hatzopoulou & Miller, 2009). A recent analysis of regional transportation plans confirms this assessment, showing fewer than 20 percent of a nationally representative sample of U.S. regional transportation plans utilizing some type of cumulative opportunity metric and none using gravity, utility, or space-time measures (Proffitt, Bartholomew, Ewing, & Miller, 2015).

Too often, assessment tools approach accessibility solely from a transportation connectivity perspective, and not from the land use proximity angle. In the U.K., the national government has been promoting accessibility planning for more than 20 years, but very few Britons live in areas with such plans (Halden, 2014). A 2013 Parliamentary Inquiry into the issue found a primary lapse in policy implementation came from the failure to incorporate accessibility factors into location decisions for services and activities, particularly those related to employment, education, and health care (House of Commons, 2013). A solution suggested by some (Curtis, 2008) is to follow the lead of the Netherlands in its use of a business location planning directive, known as the "ABC Policy."

The ABC Policy or, more formally, "The Right Business in the Right Place," was launched in 1988 as part of the Netherland's national land use planning policy (the Fourth Report on Physical Planning). The policy has at its root a very simple and compelling structure: Businesses that have a high potential for generating transit riders should be located on sites with high levels of transit service; businesses with a low potential for transit use or that require direct roadway access for freight movement should be located on sites with good road access.

Implementation of the ABC policy at local and regional levels depends on a two-pronged analysis assessing (1) the transport accessibility of lands within the urban area, and (2) the mobility profiles of various business types. Both analyses require classification of their respective subjects into three categories: A, B, or C. Sites with superior transit accessibility are classified as "A" sites, those with moderate access to transit and roadway systems are "B" sites, and those with low access to transit but good access to roadways are "C" sites. In a similar fashion, businesses with a high potential for transit ridership are classified as "A" businesses, those with a moderate potential for transit use are "B" businesses, and those with low transit potential and/or demonstrated needed for highway transport are "C" businesses. The natural object of the policy is to guide permitting decisions for new and relocating businesses so that A businesses are located on A sites, B businesses on B sites, and C businesses on C sites. Such a matching of business and location type, would, logically, increase accessibility by placing the highest number of exchange opportunities in locations with the best connections to high-capacity transit. It would also maximize the transit ridership potential of commercial land uses. This would, in turn, reduce energy consumption, greenhouse gas emissions, and air and water pollution; and promote a more socially equitable transportation system by attracting fare income and private capital that facilitate further expansion of the transit system.

An initial evaluation of the ABC policy based on conditions in The Hague estimated that A businesses in A locations could achieve 41 percent transit mode share, compared to 12 percent for A businesses in C locations (Verroen & Jansen, 1992). Ten years after implementation, most local planning authorities were reported as having integrated the ABC policy into local planning practices, although some of the restrictive parking standards associated with the policy were less rigorously incorporated (Martens & Griethuysen, 1999). Because of the slow rate of change in development patterns over time, particularly in a built-up country like the Netherlands, the policy's effects have been slow to accrue (Alpkokin, 2012). In places with rapid development like the United States, however, one would anticipate quicker results.

MPOs tend to base planning decisions on metrics that they are already generating through standard travel demand models (Handy, 2008). This reality suggests that wide adoption of accessibility as a planning framework is dependent on the use of a simple measure, one that can readily be implemented from existing processes. Gravity-based accessibility measures are generated in the trip distribution step of the standard four-step travel-demand forecasting process, but are very difficult to communicate. Among the measures available from travel-demand forecasting, and easy to communicate, vehicle miles traveled (VMT) presents itself as a possible candidate. As a performance measure, VMT has a simple elegance. "If development is compact, VMT will be low. If land uses are mixed, VMT will be low. If the road network provides direct connections, VMT will be low. If transit and ridesharing are well utilized, VMT will be low" (Ewing, 1996b).

The use of VMT as a proxy measurement for accessibility is attractive, logically. If accessibility is the measure of ease of access to exchange opportunities, it makes sense that, all things being equal, as access becomes easier, travel would decrease. This relationship is based on three basic assumptions about human behavior suggested by Bertolini, le Clercq, and Kapoen (2005):

- Travel is a derived demand, meaning people are mobile for the purpose of engaging in exchanges, not for the sake of travel itself;
- People have choices among a wide range of exchange opportunities; and
- Travel time budgets place a limit on how much travel people can consume.

These factors are evident in the few direct studies of accessibility and VMT, which show that, on balance, as accessibility increases, travel tends to decrease (Curtis, 2008). Although, accessibility can, for a period of time, increase with higher levels of mobility (faster travel speeds), a recent study of 38 U.S. metro areas shows a greater increase in accessibility from high-density development patterns, which are associated with lower travel-demand levels, than from mobility alone (Levine, Grengs, & Shen, 2012). Other studies have shown that people living in neighborhoods with high levels of accessibility areas tend to walk more and drive less than those in low-accessibility areas, and that these effects are independent of possible self-selection bias (Cao & Mokhtarian, 2012). In their meta-analysis of more than 200 land use-transportation studies, Ewing and Cervero (2010) found that among the "D" variables, VMT is most strongly related to destination accessibility (the other Ds are development density, land use diversity, street network design, and distance to transit).

These studies demonstrate VMT's satisfaction of the first criterion set out by Geurs and van Wee for accessibility measure suitability: It reflects changes in transportation capacity and land use proximity. VMT meets the remaining criteria, as well, in that it is readily understandable by the public; is capable of capturing economic, social, and ICT impacts; and, most importantly, it can be operationalized with existing data and resources by most MPOs. In fact, more than 40 percent of the MPOs in Proffitt et al.'s survey of transportation plans already use VMT as an evaluation measure (though not always for the specific purpose of assessing accessibility).

It is for these reasons that state governments have started using VMT as a central measurement tool for transportation planning. Oregon led out first with its Transportation Planning Rule, an administrative rule adopted in 1991 that directs the state's MPOs to achieve a 5 percent reduction in VMT per person over a 20-year planning horizon. Significantly, the rule calls for this reduction to be achieved largely through measures designed to improve destination proximity such as increases in density, land use mixing, and pedestrian design (Oregon Administrative Rule 660–012–035). The state's legislature has further advanced VMT reduction as part of its directive to MPOs to reduce greenhouse gas emissions (SB 1059). California's SB 375 similarly requires MPOs to reduce GHG emissions from driving, which is understood as reducing VMT. The state is also now using VMT (instead of Level of Service) to measure transportation impacts from proposed developments subject to the reporting requirements of the California Environmental Quality Act (CEQA). In Washington, the state legislature has directed the state's DOT to adopt goals to reduce VMT per person by 50 percent over 1990 levels by 2050 (RCW 47.01.440).

These are laudable – some would say daunting – goals, especially given the VMT growth trends of the 20th century. But it is not the 20th century any more. Millennials are the largest generation in the United States, which means that their choices about transportation (and just about everything else) are amplified as they cascade through the economy and society. And what they are choosing is something different from their parents'/grandparents' generations. As reported above, Millennial teenagers have been getting drivers licenses at much lower rates than previous generations. As it turns out, by the time Millennials reach 24, that difference disappears – between 82 percent and 86 percent are licensed by 24, which is roughly the equivalent level of licensure for 24-year-olds since the mid-20th century. This indicates that Millennials are delaying licensure, not forgoing it. While disappointing to some who were hoping that Millennials were forsaking cars entirely, the delay in licensure presents an excellent opportunity for policy makers to take a relatively car-disinclined demographic group and divert them to more sustainable transport modes: "Each year without a driving license extends the window of opportunity for a young person to familiarize him or herself with a range of niche innovations, from active travel to car sharing to intermodal transport" (Delbosc, 2016, p. 6). These diversions can reinforce different, more sustainable choices for accessing life's exchange opportunities – not just in whether and how to travel but also where to live and work. Over time, these different choices become different habits. The onus is on policy makers to respond effectively to this generational window and provide these new citizens with options and positive support.

The starting point for creating these options is in the transportation plans our regional leaders are making in the hundreds of metro areas across the country. This book has traced a trajectory in regional transportation planning, showing how the standard practices of building highway capacity half a century ago are giving way to new directions in planning that incorporate land use decision making, multiple travel modes, and a reorientation away from strict focus on mobility and toward accessibility.

Our communities comprise interlocking systems of interdependent elements. The arc of planning practice, as documented here, is aimed toward increasing integration of those elements. Areas that were once thought to be insular areas of life and public policy are now understood to be interconnected with a multitude of other areas. The demands on planners will only

Photo: Keith Bartholomew.

increase as this trend continues. The challenges of coming decades – climate change, energy production and consumption, economic resiliency – demand integrative approaches that rely on multiple disciplines all pulling toward achieving a set of coordinated goals. It will fall to planning to provide the connective tissue for this process, and it will be planning's challenge and opportunity to show the way toward a sustainable and resilient future.

References and further reading

1000 Friends of Oregon. (1991). *Making the land use, transportation, air quality connection: Modeling practices* (Vol. 1). Portland, OR: Author.

1000 Friends of Oregon. (1997). *Making the connection: A summary of the LUTRAQ project*. Portland, OR: 1000 Friends of Oregon.

Abdel-Aty, M. A., & Jovanis, P. P. (1995). The effect of ITS on transit ridership. ITS *Quarterly*, Fall, 21–25.

Akron Metropolitan Area Transportation Study. (2013). *Transportation outlook 2035: Regional transportation plan*. Retrieved from http://amatsplanning.org/wp-content/uploads/AMATS-Transportation-Outlook-2035.pdf

Alamo Area Metropolitan Planning Organization. (2015). *Mobility 2040*. Retrieved from http://www.alamoareampo.org/Plans/MTP/docs/Mobility2040/Final%20MTP%20Revised%20March%2010%202015.pdf

Ameli, S. H., Hamidi, S., Garfinkel-Castro, A., & Ewing, R. (2015). Do better urban design qualities lead to more walking in Salt Lake City, Utah? *Journal of Urban Design*, *20*, 393–410.

Alpkokin, P. (2012). Historical and critical review of spatial and transport planning in the Netherlands. *Land Use Policy*, *29*, 536–547.

American Association of State Highway and Transportation Officials (AASHTO). (2012). *Guide for the planning, design, and operation of bicycle facilities*. Washington, DC: American Association of State Highway and Transportation Officials. Retrieved from http://design.transportation.org/Documents/DraftBikeGuideFeb2010.pdf

American Public Transit Association. (APTA). (2009), *Quantifying greenhouse gas emissions from transit*. Washington, D.C.: APTA.

American Society of Civil Engineers (ASCE). (2010). *Report card for America's infrastructure*. Retrieved from www.infrastructurereportcard.org/

Anchorage Metropolitan Area Transportation Solutions. (2012). *2035 metropolitan transportation plan*. Retrieved from http://kaiproject.com/system/images/650/original/Anchorage.pdf

Anderson, J. (2003). *Public policy-making*. New York: Praeger.

Anderson, P., Levinson, D., & Parthasarathi, P. (2013, May). Accessibility futures. *Transactions in GIS*, *17*(5).

Apogee Research, Inc., & Hagler Bailly. (1998). *The effects of urban form on travel and emissions: A review and synthesis of the literature*. Washington, DC: U.S. Environmental Protection Agency.

AP Staff. (2017, February 3). EPA: Oklahoma regulators should do more to curb earthquakes. *The Associated Press*. Retrieved from http://bigstory.ap.org/article/aada3565b87a4b90a908bfa1e58b3942/epa-oklahoma-regulators-should-do-more-curb-earthquakes

Arline, K. (2014, December 11). What is a vision statement? *Business News Daily*.

Arnstein, S. R. (1969). A ladder of citizen participation. *Journal of the American Institute of Planners*, *35*(4), 216–224.

Association of Central Oklahoma Governments. (2012). *Encompass 2035: Oklahoma City regional transportation study*. Retrieved from http://www.acogok.org/wp-content/uploads/2015/09/2035_Plan_Report.pdf

Association for Commuter Transportation, Urbantrans Consultants, PB Parsons Brinckerhoff, & ESTC. (2004). *Mitigating traffic congestion: The role of demand-side strategies*. Federal Highway Administration and the U.S. Department of Transportation. Retrieved from www.ops.fhwa.dot.gov/publications/mitig_traf_cong/

Atkinson-Palombo, C. (2010). Comparing the capitalization benefits of light-rail transit and overlay zoning for single-family houses and condos by Neighborhood type in Metropolitan Phoenix, Arizona. *Urban Studies*, *47*, 2409–2426.

Atlanta Regional Commission. (2008). *2030 regional transportation plan*. Retrieved from www.atlantaregional.com/transportation/regional-transportation-plan

Atlanta Regional Commission (2014). *PLAN 2040 regional transportation plan (RTP)* Retrieved from http://documents.atlantaregional.com/plan2040/docs/tp_PLAN2040RTP_072711.pdf

Augusta-Richmond County Planning Commission. (2010). *Augusta regional transportation study: 2035 long range transportation plan*. Retrieved from http://appweb.augustaga.gov/Transporation/docs/ARTS2035LRTP_Final.pdf

Badland, H., & Schofield, G. (2005). Transport, urban design, and physical activity: An evidence-based update. *Transportation Research D*, *10*(3), 177–196.

Badoe, D. A., & Miller, E. J. (2000). Transportation: Land-use interaction: Empirical findings in North America, and the implications for modeling. *Transportation Research D*, *5*(4), 235–263.

Bailey, L., Mokhtarian, P., & Little, A. (2008). *Broadening the connection between public transportation and energy conservation*. Fairfax, VA: ICF Consulting.

Bain, D. H. (1999). *Empire express: Building the first transcontinental railroad*. New York: Penguin.

Bakker, J. J., et al. (1988). Multi-centered time transfer system for Capital Metro, Austin, Texas. *Transportation Research Record*, *1202*, 22–28.

Baltimore Regional Transportation Board. (2007). *Transportation outlook 2035*. Retrieved from www.baltometro.org/downloadables/Outlook2035/

Baltimore Regional Transportation Board. (2011). *Plan it 2035*. Retrieved from https://www.baltometro.org/our-work/transportation-plans/long-range-planning/plan-it-2035

Barnes, G., Thompson, K., & Krizek, K. J. (2006). *A longitudinal analysis of the effect of bicycle facilities on commute mode share*. TRB 85th Annual Meeting Compendium of Papers CD-ROM. Washington, DC: Transportation Research Board.

Bartholomew, K. (1995). A tale of two cities. *Transportation*, *22*, 273–293.

Bartholomew, K. (2005). *Integrating land use issues into transportation planning: Scenario planning* (bibliography). Salt Lake City, UT: University of Utah.

Bartholomew, K. (2007a). Land use-transportation scenario planning: Promise and reality. *Transportation*, *34*(4), 397–412.

Bartholomew, K. (2007b). The machine, the garden, and the city: Toward an access-efficient transportation planning system. *Environmental Law Reporter*, *37*(8), 10593–10614.

Bartholomew, K. (2009). Cities and accessibility: The potential for carbon reductions and the need for national leadership. Fordham *Urban Law Journal*, *36*(2), 159–209.

Bartholomew, K., & Ewing, R. (2008). Land use-transportation scenarios and future vehicle travel and land consumption: A meta-analysis. *Journal of the American Planning Association*, *75*(1), 1–15.

Bartholomew, K., & Ewing, R. (2009). Land use-transportation scenario planning: A meta-analysis. *Journal of the American Planning Association*, *75*(1), 1–15.

Bartholomew, K., & Ewing, R. (2011). The hedonic price effects of pedestrian-and transit-oriented development. *Journal of Planning Literature*, *26*(1), 18–34.

Baum-Snow, N., & Kahn, M. E. (2005). *The effects of urban rail transit expansions: Evidence from sixteen cities, 1970 to 2000*. Washington, DC: Brookings Institution. Retrieved March 11, 2011 from www.econ.brown.edu/fac/Nathaniel_Baum-Snow/brook_final.pdf

Bauman, A. E., & Bull, F. C. (2007). *Environmental correlates of physical activity and walking in adults and children: A review of reviews*. London, U.K.: National Institute of Health and Clinical Excellence.

Beimborn, E., Kennedy, R., & Schaefer, W. (n.d.). *Inside the blackbox: Making transportation models work for livable communities*. New York: Environmental Defense Fund.

Belden Russonello, & Stewart. (2004). *National survey on communities*. Washington, DC: National Association of Realtors and Smart Growth America. Retrieved from www.brspoll.com/Reports/Smart%20Growth.pdf

Bellah, R. N., Madsen, R., Sullivan, W. M., Swidler, A., & Tipton, S. M. (1992). *The good society*. New York: Vintage Books.

Bento, A. M., Cropper, M. L., Mobarak, A. M., & Vinha, K. (2003). *The impact of urban spatial structure on travel demand in the united states*. World Bank Group Working Paper 2007. Washington, DC: World Bank. Retrieved March 11, 2011 from http://papers.ssrn.com/sol3/papers.cfm?abstract_id=636369

Berkeley-Charleston-Dorchester Council of Governments. (2010). *2035 CHATS: Long range transportation plan*. Retrieved from https://bcdcog.com/long-range-transportation-plan/

Berks County Planning Commission. (2014). *Reading area transportation study: Long range transportation plan*. Retrieved from http://www.co.berks.pa.us/Dept/Planning/Documents/Transportation/Final_LRP/LRTP_15-40_FINAL_Complete.pdf

Berkshire Regional Planning Commission. (2012). *Berkshire regional transportation plan*. Retrieved from http://berkshireplanning.org/images/uploads/documents/Regional_Transportation_Plan_2011.pdf

Bertolini, L., le Clercq, F., & Kapoen, L. (2005). Sustainable accessibility: A conceptual framework to integrate transport and land use plan-making: Two test-applications in the Netherlands and a reflection on the way forward. *Transport Policy*, *12*(3), 207–220.

Bhatta, B., & Larsen, O. I. (2011). Are intrazonal trips ignorable? *Transport Policy*, *18*(1), 13–22.

Bismarck-Manden Metropolitan Planning Organization. (2009). *2009–2035 long range transportation plan*. Retrieved from http://bis-manplan2009.com/PDFs/Final%20LRTP_Web.pdf

Bismarck-Mandan Metropolitan Planning Organization. (2010). *2010-2035 long range transportation plan report*. Retrieved from http://www.bismarcknd.gov/DocumentCenter/View/4139

Boston Region Metropolitan Planning Organization. (2007). *Journey to 2030*. Retrieved from www.ctps.org/bostonmpo/3_programs/1_transportation_plan/plan.html

Boston Region Metropolitan Planning Organization. (2011). *Paths to a sustainable region 2035*. Retrieved from http://www.adaptationclearinghouse.org/resources/paths-to-a-sustainable-region-boston-region-long-range-transportation-plan.html

Boston Region Metropolitan Planning Organization. (2015). Charting progress to 2040: Long range transportation plan of the Boston Region Metropolitan Planning Organization. Retrieved from http://www.ctps.org/data/pdf/plans/lrtp/charting/2040_LRTP_Full_final.pdf

Bradley, M. S. (1993). *Sandy city: The first 100 years*. Sandy, UT: Sandy City Corp.

Brandford, V., & Geddes, P. (1919). *The coming polity*. London: Williams and Norgate.

Brons, M., Nijkamp, P., Pels, E., & Rietveld, P. (2006). *A meta-analysis of the price elasticity of gasoline demand: A system of equations approach*. Amsterdam: Tinbergen Institute Discussion Paper.

Broward Metropolitan Planning Organization. (2009). *2035 Broward transformation: Long range transportation plan*. Retrieved from http://www.browardmpo.org/index.php/33-long-range-transportation-plan-lrtp

Brown, J., Nixon, H., & Ramos, E. (2015). *The Purpose, Function, and Performance of Streetcar Transit in the Modern US City: A Multiple-Case-Study Investigation* (No. CA-MTI-14-1201). San Jose, CA: Mineta Transportation Institute.

Brown, J., & Thompson, G. (2008). Service orientation, bus—rail service integration, and transit performance: Examination of 45 US metropolitan areas. *Transportation Research Record: Journal of the Transportation Research Board*, *2042*, 82–89.

Brown, J., & Thompson, G. (2009). Express bus versus rail transit: How a marriage of mode and mission affects transit performance. *Transportation Research Record: Journal of the Transportation Research Board*, *2110*, 45–54.

Brown, J., Thompson, G., Bhattacharya, T., & Jaroszynski, M. (2014). Understanding transit ridership demand for the multidestination, multimodal transit network in Atlanta, Georgia: Lessons for increasing rail transit choice ridership while maintaining transit dependent bus ridership. Urban *Studies*, *51*(5), 938–958.

Brundtland, G. H. (1987). *Report of the World Commission on environment and development: our common future*. United Nations.

Buehler, R., Hamre, A., Sonenklar, D., & Goger, P. (2009). *Trends and determinants of cycling in the Washington, DC region* (USDOT Report No. VT-2009–05). Alexandria, VA: USDOT Research and Technology Administration.

Buehler, R., & Pucher, J. (2012). Cycling to work in 90 large American cities: New evidence on the role of bike paths and lanes. *Transportation*, *39*, 409–432.

Bullard, R. D., Johnson, G. S., & Torres, A. O. (2001). The costs and consequences of suburban sprawl: The case of metro Atlanta. *Georgia State University Law Review*, *17*, 935–997.

Burchell, R., et al. (2002). *Costs of sprawl – 2000*. Washington, DC: National Academies Press.

Bureau of Public Roads. (1963). *Urban transportation planning: Instructional memorandum number 50–2–63*. Washington, DC: U.S. Department of Commerce.

Butler, K., Handy, S., & Paterson, R. G. (2003). *Planning for street connectivity* (Planning Advisory Service Report 515). Chicago, IL: American Planning Association.

California Air Resources Board. (1991). *Transportation performance standards of the California Clean Air Act*. Sacramento, CA: California Air Resources Board, p. 3.

Calthorpe, P., & Fulton, W. (2001). *The regional city: Planning for the end of Sprawl*. Washington, DC: Island Press.

Cambridge Systematics, Inc. (2015). *Status of activity-based models and dynamic traffic assignment at peer MPOs*. Retrieved from www.mwcog.org/transportation/data-and-tools/modeling/review-of-travel-modeling-procedures/

Cao, X., & Mokhtarian, P. L. (2012). The connections among accessibility, self-selection and walking behaviour: A case study of Northern California residents. In K. T. Geurs, K. J. Krizek, & A. Reggiani (Eds.), *Accessibility analysis and transport planning*. Cheltenham, UK: Edward Elgar Publishing.

Capital Area Metropolitan Planning Organization, & Durham-Chapel Hill-Carrboro Metropolitan Planning Organization. (2009). *2035 long range transportation plans*. Retrieved from http://ourtransitfuture.com/wp-content/uploads/2013/12/3-00627-DCHC-2035-LRTP-June19-Final.pdf

Capital Area Metropolitan Planning Organization and Durham-Chapel Hill-Carrboro Metropolitan Planning Organization. (2013). *2040 long range transportation plans*. Retrieved from www.campo-nc.us/lrtp.html

Capital Area Metropolitan Planning Organization. (2010). *CAMPO 2035 regional transportation plan*. Retrieved from http://www.campotexas.org/wp-content/uploads/2013/10/CAMPO_2035_Plan_Adopted_May_242010wMods.pdf

Capital District Transportation Committee (Albany, NY). (2004). *New visions 2025 amendment*. Retrieved from www.cdtcmpo.org/rtp2025/2025.htm

Capital District Transportation Committee. (2007). *New Visions 2030: The Plan for a Quality Region*. Summary Document. Albany, NY: Capital District Transportation Committee.

Capital District Transportation Committee. (2011). *Choosing our future: New visions 2035 plan update*. Retrieved from http://www.cdtcmpo.org/rtp2035/summary.pdf

Capital Region Council of Governments. (2011). *Capital Region transportation plan*. Retrieved from http://crcog.org/wp-content/uploads/2016/05/2011-RTPFinal.pdf

Capital Region Planning Commission. (2013). *Baton Rouge metropolitan transportation plan 2037*. Retrieved from https://crpc.sharefile.com/share/view/2d13144a9e464d1b

Capital Region Transportation Planning Agency. (2006). *Capital region transportation planning agency year 2030 long range transportation plan: Cost feasible plan*. Retrieved from www.crtpa.org/

Capital Region Transportation Planning Agency. (2010). *Regional Mobility Plan. 2035 Long Range Transportation Plan*.

Capital Region Transportation Planning Agency. (2015). *Connecting people and places: Regional mobility plan*. Retrieved from http://www.crtpa.org/connections-2040.html

Carbonell, A., & Yaro, R. D. (2005). American spatial development and new megalopolis. *Land Lines*, *17*(2), 1–4.

Castiglione, J., Bradley, M., & Gliebe, J. (2015). Activity-based travel demand models: A primer. *Transportation Research Board*. Retrieved from http://onlinepubs.trb.org/onlinepubs/shrp2/SHRP2_C46.pdf

Center for Urban Transportation Research (CUTR). (2004). *Worksite trip reduction model and manual*. Tampa: University of South Florida. Retrieved from www.dot.state.fl.us/research-center/Completed.../FDOT_BC137_41.pdf

Central Lane Metropolitan Planning Organization. (2007). *Regional transportation plan*. Retrieved from www.thempo.org/prog_proj/rtp.cfm

Central Lane Metropolitan Planning Organization. (2011). *Regional transportation plan*. Retrieved from http://www.lcog.org/DocumentCenter/View/670

Central Massachusetts Metropolitan Planning Organization. (2012). *Congestion management process: Progress report*. Retrieved from http://www.cmrpc.org/sites/default/files/Documents/Trans/Study_and_Plan/CMP/2012%20CMP%20Report.pdf

Cervero, R. (1988). Land use mixing and suburban mobility. *Transportation Quarterly*, *42*, 429–446.

Cervero, R. (1989a). *America's suburban centers: The land use-transportation link*. Boston: Unwin Hyman.

Cervero, R. (1989b). Jobs-housing balancing and regional mobility. *Journal of the American Planning Association*, *55*, 136–150.

Cervero, R. (1991). Land uses and travel at suburban activity centers. *Transportation Quarterly*, *45*(4), 479–491.

Cervero, R. (1993). *Ridership impacts of transit-focused development in California*. Berkeley, CA: Institute of Urban and Regional Development, University of California.

Cervero, R. (1994a). Rail-oriented office development in California: How successful? *Transportation Quarterly*, *48*(1), 33–44.

Cervero, R. (1994b). Transit-based housing in California: Evidence on ridership impacts. *Transport Policy*, *1*(3), 174–183.

Cervero, R. (1996a). Mixed land-uses and commuting: Evidence from the American Housing Survey. *Transportation Research Part A*, *30*(5), 361–377.

Cervero, R. (1996b). Jobs-housing balance revisited: Trends and impacts in the San Francisco Bay Area. *Journal of the American Planning Association*, *62*, 492–510.

Cervero, R. (2001). Walk-and-ride: Factors influencing pedestrian access to transit. *Journal of Public Transportation*, *3*(4), 1–23.

Cervero, R. (2002). Induced travel demand: Research design, empirical evidence, and normative policies. *Journal of Planning Literature*, *17*(1), 3–20.

Cervero, R. (2003). The built environment and travel: Evidence from the United States. *European Journal of Transport and Infrastructure Research*, *3*(2), 119–137.

Cervero, R. (2006). Alternative approaches to modeling the travel-demand impacts of smart growth. *Journal of the American Planning Association*, *72*(3), 285–295.

Cervero, R. (2007). Transit oriented development's ridership bonus: A product of self-selection and public policies. *Environment and Planning Part A*, *39*, 2068–2085.

Cervero, R., & Duncan, M. (2002). *Residential self-selection and rail commuting: A nested logit analysis*. Working Paper. Berkeley, CA: University of California Transportation Center.

Cervero, R., & Duncan, M. (2003). Walking, bicycling, and urban landscapes: Evidence from the San Francisco Bay Area. *American Journal of Public Health, 93*(9), 1478–1483.

Cervero, R., & Duncan, M. (2006). Which reduces vehicle travel more: Jobs-housing balance or retail-housing mixing. *Journal of the American Planning Association, 72*(4), 475–490.

Cervero, R., & Gorham, R. (1995). Commuting in transit versus automobile neighborhoods. *Journal of the American Planning Association, 61*(2), 210–225.

Cervero, R., & Griesenbeck, B. (1988). Factors influencing commuting choices in suburban labor markets: A case analysis of Pleasanton, California. *Transportation Research A, 22*(3), 151–161.

Cervero, R., & Kockelman, K. (1997). Travel demand and the 3Ds: Density, diversity, and design. *Transportation Research D, 2*(3), 199–219.

Cervero, R., & Landis, J. (1995). The transportation-land use connection still matters. *Access, 7*, 2–10.

Cervero, R., & Murakami, J. (2010). Effects of built environments on vehicle miles traveled: Evidence from 370 U.S. metropolitan areas. *Environment and Planning A, 42*, 2.

Cervero, R., Murphy, S., Ferrell, C., Goguts, N., & Tsai, Y. (2004).*Transit-oriented development in the United States: Experiences, challenges, and prospects* (Report No. 102). Washington, DC: Transit Cooperative Research Program.

Cervero, R., & Radisch, C. (1996). Travel choices in pedestrian versus automobile oriented neighborhoods. *Transport Policy, 3*(3), 127–141.

Cervero, R., Sarmiento, O., Jacoby, E., Gomez, L., & Neiman, A. (2009). Influences of built environments on walking and cycling: Lessons from Bogotá. *International Journal of Sustainable Transport, 3*, 203–226.

Cervero, R., & Seskin, S. (1995). *An evaluation of the relationships between transit and urban form* (Research Results Digest). Washington, DC: Transit Cooperative Research Program, Transportation Research Board.

Charlotte Regional Transportation Planning Organization. (2014). *Metropolitan transportation plan 2040*. Retrieved from http://crtpo.org/PDFs/MTP/2040/Report/CRTPO_2040MTP.pdf

Charlottesville/Albemarle Metropolitan Planning Organization. (2004). UnJAM 2025 United Jefferson area mobility plan.

Charlottesville/Albemarle Metropolitan Planning Organization. (2014). *Long range transportation plan 2040*. Retrieved from http://campo.tjpdc.org/process-documents/2040-lrtp/

Chatham County-Savannah Metropolitan Planning Commission. (2009). *2035 core connections*. Retrieved from www.thempc.org/documents/Transportation/2035%20LRTP/Final/Complete%20Document.pdf

Chattanooga Hamilton County North Georgia Transportation Planning Organization. (2009). *Chattanooga Hamilton County North Georgia 2030 long range transportation plan*. Retrieved from www.chcrpa.org/TPO/.../LRTP/LRTP_2030/lrtp/Section_5_%20Fin_Res.pdf

Chattanooga-Hamilton County Regional Planning Agency. (2013). *North Georgia 2040 regional transportation plan*. Retrieved from https://drive.google.com/file/d/1-Mc3eaZnCgSFsZ0nxuonChEXxpCQNUuR/view

Chen, C., & McKnight, C. E. (2007). Does the built environment make a difference? Additional evidence for the daily activity and travel behavior of homemakers living in New York City and suburb. *Journal of Transport Geography*, *15*(5), 380–395.

Chicago Department of Transportation. (2012). *Protected bike lanes overview*. Retrieved from www.chicagobikes.org/pdf/Cycle_Tracks_Overview.pdf

Chicago Metropolitan Agency for Planning. (2008). *Updated 2030 regional transportation plan for Northeastern Illinois*. Retrieved from www.cmap.illinois.gov/sp2030/sp2030main.aspx?ekmensel=c580fa7b_8_12_5564_6

Chicago Metropolitan Agency for Planning. (2010). *Go to 2040: Comprehensive regional plan*. Retrieved from http://www.cmap.illinois.gov/documents/10180/17842/long_plan_FINAL_100610_web.pdf/1e1ff482-7013-4f5f-90d5-90d395087a53

Chittenden County Metropolitan Planning Organization. (2005). *2025 Chittenden County metropolitan transportation plan*. Retrieved from www.ccmpo.org/MTP/2025/

Chittenden County Regional Planning Commission. (2013). *ECOS plan*. Retrieved from https://www.ccrpcvt.org/wp-content/uploads/2016/01/ECOS-Plan-Final-MERGED-20160610.pdf

City of Copenhagen. (2011). *Copenhagen bicycle account*. Copenhagen: Traffic Department, The Technical and Environmental Administration.

Clean Air Act (CAA) Amendments of 1990, 42 U.S.C. §§7401–7671q.

Cleveland Urban Area Metropolitan Planning Organization. (2006). *2030 long range transportation plan*. Retrieved from www.cityofclevelandtn.com/MPO/mpolrtp.html

Cleveland Urban Area Metropolitan Planning Organization. (2011). *2035 regional transportation plan*. Retrieved from http://www.clevelandtn.gov/DocumentCenter/View/195

Coastal Region Metropolitan Planning Organization. (2009). *CORE connections 2035*. Retrieved from http://www.thempc.org/docs/lit/CoreMpo/Plans/TotalMobility/2009/Sep/FinalDoc.pdf

Coleman, S., & Gotze, J. (2001). *Bowling together: Online public engagement in policy deliberation*. London: Hansard Society.

Cobian, R., Henderson, T., Mitra, S., Nuworsoo, C., & Sullivan, E. (2009). Vehicle emissions and level of service standards: Exploratory analysis of the effects of traffic flow on vehicle greenhouse gas emissions. *ITE J*, *79*(4), 30–41.

Columbia Area Transportation Study. (2007). *Midlands tomorrow: 2035 long range transportation plan*. Retrieved from http://www.centralmidlands.org/COATS_2035_pdf/COATS_FinalPlan_Rev_5-11-10.pdf

Committee for Determination of the State of the Practice in Metropolitan Area Travel Forecasting. (2007). *Metropolitan travel forecasting: Current practice and future direction* (Special Report 288). Washington, DC: Transportation Research Board.

Committee on Access Management. (2003). *Access management manual*. Washington, DC: Transportation Research Board.

Committee on Physical Activity, Health, Transportation, and Land Use of the Transportation Research Board Institute of Medicine (2005). *Does the Built Environment Influence Physical Activity? Examining the Evidence*. Washington DC. Retrieved from http://onlinepubs.trb.org/onlinepubs/sr/sr282.pdf

Community Planning Association of Southwest Idaho. (2006). *Communities in motion: Regional long-range transportation plan 2030*. Retrieved from www.communitiesinmotion.org/plandocuments.html

Compass Community Planning Association of Southwest Idaho. (2010). *Communities in motion 2035: Regional long-range transportation plan*. Retrieved from http://www.compassidaho.org/documents/prodserv/CIMupdate/2010/FINAL/Communities_in_Motion_Entire_Plan.pdf

Contra Costa Transportation Authority. (2014). *Countywide comprehensive transportation plan*. Retrieved from http://ccta.net/uploads/53ebd3bbe4fcb.pdf

Crane, R. (2000). The influence of urban form on travel: An interpretive review. *Journal of Planning Literature*, *15*(1), 3–23.

Cunningham, G. O., & Michael, Y. L. (2004). Concepts guiding the study of the impact of the built environment on physical activity for older adults: A review of the literature. *American Journal of Health Promotion*, *18*(6), 435–443.

Curl, A., Nelson, J. D., & Anable, J. (2011). *Does accessibility planning address what matters? A review of current practice and practitioner perspectives* (Research in Transportation Business & Management). Glasgow, UK: University of Glasgow.

Currie, G., & Loader, C. (2009). High ridership growth from extended transit service hours: An exploration of the causes. Transportation *Research Record: Journal of the Transportation Research Board*, *2110*, 120–127.

Curtis, C. (2008). Planning for sustainable accessibility: The implementation challenge. *Transport Policy*, *15*, 104–112.

Danaher, A., Levinson, H. S., & Zimmerman, S. L. (2007). *TCRP report 118: Bus rapid transit practitioner's guide* (Transportation Research Board of the National Academies). Washington, DC: Federal Transit Administration.

Davison, K. K., & Lawson, C. T. (2006). Do attributes in the physical environment influence children's physical activity? A review of the literature. *The International Journal of Behavioral Nutrition and Physical Activity*, *3*, 19. http://doi.org/10.1186/1479-5868-3-19

DeFilippis, J. (2008). Community control and development. In J. DeFilippis & S. Saegert (Eds.), *The community development reader* (Chapter 3). New York and London: Routledge.

de Graaf, T., Ghebreegziabiher, D., & Rietveld, P. (2012). The impact of accessibility on house prices: An application to large urban planning and infrastructure projects in the Netherlands. In K. T. Geurs, K. J. Krizek, & A. Reggiani (Eds.), *Accessibility analysis and transport planning*. Cheltenham, UK: Edward Elgar Publishing.

Delaware Valley Regional Planning Council. (2009). *Connections 2035: The regional plan for a sustainable future: The long-range plan for the greater Philadelphia region*. Retrieved from www.dvrpc.org/asp/pubs/publicationabstract.asp?pub_id=09047

Delaware Valley Regional Planning Commission. (2013). *Connections 2040: Plan for Greater Philadelphia*. Retrieved from https://www.dvrpc.org/reports/13042.pdf

Delbosc, A. (2016, February 24). Delay or forgo? A close look at youth driver licensing trends in the United States and Australia. *Transportation*. DOI: 10.1007/s11116-016-9685-7

Delbosc, A., & Currie, G. (2013). Causes of youth licensing decline: A synthesis of evidence. *Transport Reviews*, *33*(3), 271–290.

Denver Regional Council of Governments. (2011). *2035 metro vision regional transportation plan*. Retrieved from https://drcog.org/planning-great-region/metro-vision/metro-vision-2035

Denver Regional Council of Governments. (2017). *2040 metro vision regional transportation plan*. Retrieved from https://drcog.org/sites/drcog/files/resources/

FINAL%20-%202040%20MVRTP%20w%20APPENDICES%20-%20April%202017.pdf
Des Moines Area Metropolitan Planning Organization. (2009). *Horizon year 2035 metropolitan transportation plan*. Retrieved from https://dmampo.org/regional-plans/
Diamond, J. (1999). *Guns, germs, and steel: The fates of human societies*. New York: W. W. Norton.
DiGuilio, D., & Jackson, R. (2016). Impact to underground sources of drinking water and domestic wells from production well stimulation and completion practices in the Pavillion, Wyoming field. *Environmental Science and Technology, 50*(8), 4524–4536.
Dill, J. (2009). Bicycling for transportation and health: the role of infrastructure. *Journal of Public Health Policy, 30*(1), S95–S110.
Dill, J. (2012). *Categorizing cyclists: What do we know? Insights from Portland, OR*. (Powerpoint Slides). Retrieved from http://web.pdx.edu/~jdill/Dill_VeloCity_Types_of_Cyclists.pdf
Dill, J., & Carr, T., (2003). Bicycle commuting and facilities in major U.S. cities: If you build them, commuters will use them. *Transportation Research Record, 1828*, 116–123.
Ding, D. & Gebel, K. (2012). Built environment, physical activity, and obesity: what have we learned from reviewing the literature? *Health Place, 18*(1): 100–105.
District Department of Transportation. (2011). *Bicycle facility evaluation executive summary*. Retrieved from http://dc.gov/DC/DDOT/Publication%20Files/On%20Your%20Street/Bicycles%20and%20Pedestrians/Bicycles/Bike%20Lanes/DDOT_BicycleFacilityEvaluation_ExecSummary.pdf
DKS Associates, & University of California. (2007). *Assessment of local models and tools for analyzing smart-growth strategies*. Irvine, CA: University of California. Retrieved from www.dot.ca.gov/newtech/researchreports/reports/2007/local_models_tools.pdf
Dover/Kent Metropolitan Planning Organization. (2005). *Moving forward together: A long range transportation plan for 2030*. Retrieved from www.doverkentmpo.org/MPOLRTP.php
Dover/Kent County Metropolitan Planning Organization. (2013). *Metropolitan transportation plan: 2040 update*. Retrieved from https://doverkentmpo.delaware.gov/files/2015/06/MTP-2040-Jan-10-2013.pdf
Dunphy, R. T., & Fisher, K. (1996). Transportation, congestion, and density: New insights. *Transportation Research Record, 1552*, 89–96.
Duthie, J., Brady, J. F., Mills, A. F., & Machemehl, R. B. (2009). Effects of on-street bicycle facility configuration on bicyclist and motorist behavior. *Transportation Research Record, 2190*, 37–44.
Eastgate Regional Council of Governments. (2013). *2040 metropolitan transportation plan*. Retrieved from http://www.eastgatecog.org/transportation/2040-metro
East-West Gateway Council of Governments. (2007). *Legacy 2035: The transportation plan for the gateway region*. Retrieved from www.ewgateway.org/trans/longrgplan/longrgplan.htm
East-West Gateway Council of Governments. (2011). *Regional transportation plan 2040*. Retrieved from http://www.ewgateway.org/transportation-planning/long-range-planning/

Edwards, J. L., & Schofer, J. L. (1976). Relationships between transportation energy consumption and urban structure: Results of simulation studies. *Transportation Research Record*, *599*, 52–59.

Ehrlich, D. (1996). *Charging for parking in suburban areas: Case studies of worksites in King County*. Paper presented at the 75th Annual Meeting of the Transportation Research Board. Washington, DC.

El Paso Metropolitan Planning Organization. (2013). *Horizon 2040: Metropolitan transportation plan*. Retrieved from http://www.elpasompo.org/civicax/filebank/blobdload.aspx?BlobID=22816

Engwicht, D. (1993). *Reclaiming our cities and towns: Better living with less traffic*. Gabriola Island, BC: New Society.

Espey, M. (1998). Gasoline demand revisited: An international meta-analysis of elasticities. *Energy Economics*, *20*, 273–295.

Evans, J. E., et al. (2004). *Traveler response to transportation system changes: Transit scheduling and frequency* (Transit Cooperative Research Program TCRP Report 95; Chapter 9). Washington, DC: Transportation Research Board.

Everett, M. D., & Spencer, J. (1983). Empirical evidence on determinants of mass bicycle commuting in the United States: A cross-community analysis. *Transportation Research Record*, *912*, 28–37.

Ewing, R. (1992a). *Beyond speed: The next generation of transportation service standards*. ITE 1992 Compendium of Technical Papers (p. 341–345). Washington, DC: Institute of Transportation Engineers.

Ewing, R. (1992b). Roadway levels of service in an era of growth management. *Transportation Research Record*, *1364*, 63–70.

Ewing, R. (1994). Residential street design: Do the British and Australians know something we Americans don't? *Transportation Research Record*, *1455*, 42–49.

Ewing, R. (1995). Beyond density, mode choice, and single-purpose trips. *Transportation Quarterly*, *49*, 15–24.

Ewing, R. (1996). Beyond speed: The next generation of transportation performance measures. In D. Porter (Ed.), *Performance standards for growth management* (pp. 31–40). Chicago: American Planning Association.

Ewing, R. (1996a). *Pedestrian-and transit-friendly design*. Tallahassee, FL: Florida Department of Transportation.

Ewing, R. (1996b). *Best development practices*. Chicago, IL: American Planning Association Press (in cooperation with the Urban Land Institute).

Ewing, R. (1997). *Transportation and land use innovations: When you can't pave your way out of congestion*. Orlando, FL: Florida Department of Community Affairs.

Ewing, R. (1999). *Traffic calming state-of-the-practice*. Washington, DC: Institute of Transportation Engineers.

Ewing, R. (2000). Sketch planning a street network. *Transportation Research Record*, *1722*, 75–79.

Ewing, R. (2001). Impacts of traffic calming. *Transportation Quarterly*, *55*(1), 33–45.

Ewing, R. (2002). Impediments to context-sensitive main street design. *Transportation Quarterly*, *56*, 51–64.

Ewing, R. (2008). Highway induced development: Results for metropolitan areas. *Transportation Research Record*, *2067*, 101–109.

Ewing, R., Bartholomew, K., Winkelman, S., Walters, J., & Chen, D. (2008). *Growing cooler: The evidence on urban development and climate change*. Washington, DC: Urban Land Institute.

Ewing, R., & Cervero, R. (2001). Travel and the built environment: A synthesis. *Transportation Research Record, 1780*, 87–114.

Ewing, R., & Cervero, R. (2010). Travel and the built environment: A meta-analysis. *Journal of the American Planning Association, 76*(3), 265–294.

Ewing, R., & Cervero, R. (2017). "Does compact development make people drive less?" The answer is yes. *Journal of the American Planning Association, 83*(1), 19–25.

Ewing, R., & Clemente, O. (2013). *Measuring urban design: Metrics for livable places*. Washington, DC: Island Press.

Ewing, R., DeAnna, M., & Li, S. (1996). Land use impacts on trip generation rates. *Transportation Research Record, 1518*, 1–7.

Ewing, R., Greenwald, M. J., & Zhang, M. (2011). School trips: Analysis of factors affecting mode choice in three metropolitan areas. In *School siting and healthy communities: Why where we invest in school facilities matters*. East Lansing, MI: Michigan State University Press.

Ewing, R., Greenwald, M. J., Zhang, M., et al. (2009). *Measuring the impact of urban form and transit access on mixed use site trip generation rates: Portland pilot study*. Washington, DC: U.S. Environmental Protection Agency.

Ewing, R., Greenwald, M. J., Zhang, M., Walters, J., Feldman, M., Cervero, R., & Thomas, J. (2011). Traffic generated by mixed-use developments: Six-region study using consistent built environmental measures. *Journal of Urban Planning and Development, 137*(3), 248–261.

Ewing, R., Greenwald, M. J., Zhang, M., Bogaerts, M., & Greene, W. (2013). Predicting transportation outcomes for LEED projects. *Journal of Planning Education and Research, 33*(3), 265–279.

Ewing, R., Haliyur, P., & Page, G. W. (1994). Getting around a traditional city, a suburban PUD, and everything in-between. *Transportation Research Record, 1466*, 53–62.

Ewing, R., & Hamidi, S. (2014). Longitudinal analysis of transit's land use multiplier in Portland (OR). *Journal of the American Planning Association, 80*(2), 123–137.

Ewing, R., Hamidi, S., Gallivan, F., Nelson, A. C., & Grace, J. B. (2014). Structural equation models of VMT growth in US urbanised areas. *Urban Studies, 51*(14), 3079–3096.

Ewing, R., & Handy, S. (2008). Measuring the unmeasurable: Urban design qualities related to walkability. *Journal of Urban Design*, 14(1), 65–84.

Ewing, R., Handy, S., Brownson, R., Clemente, O., & Winston, E. (2006). Identifying and measuring urban design qualities related to walkability. *Journal of Physical Activity and Health, 3*, s223–240.

Ewing, R., Nelson, A. C., & Bartholomew, K. (2011). Compactness vs. sprawl: Have energy, climate, health, and demographics resolved the debate? In *Urban design: Roots, influences, and trends* (The Routledge Companion to Urban Design). New York: Routledge.

Ewing, R., Nelson, A. C., Bartholomew, K., Emmi, P., & Appleyard, B. (2011). Response to special report 298: Driving and the built environment: The effects of compact development on motorized travel, energy use, and CO2 emissions. *Journal of Urbanism, 4*(1), 1–5.

Ewing, R., Pendall, R., & Chen, D. (2002). *Measuring sprawl and its impact*. Washington, DC: Smart Growth America and the U.S. Environmental Protection Agency.

Ewing, R., Pendall, R., & Chen, D. (2003). Measuring sprawl and its transportation impacts. *Transportation Research Record*, *1832*, 175–183.

Ewing, R., & Rong, F. (2008). The impact of urban form on U.S. residential energy use. *Housing Policy Debate*, *19*(1).

Ewing, R., Schieber, R., & Zegeer, C. V. (2003). Urban sprawl as a risk factor in motor vehicle occupant and pedestrian fatalities. *American Journal of Public Health*, *93*, 1541–1545.

Ewing, R., Schmid, T., Killingsworth, R., Zlot, A., & Raudenbush, S. (2003). Relationship between urban sprawl and physical activity, obesity, and morbidity. *American Journal Health Promotion*, *18*(1), 47–57.

Ewing, R., Schroeer, W., & Greene, W. (2003). *School location and student travel: Analysis of factors affecting mode choice*. Washington, DC: U.S. Environmental Protection Agency, Development, Community and Environment Division.

Ewing, R., Schroeer, W., & Greene, W. (2004). School location and student travel: Analysis of factors affecting mode choice. *Transportation Research Record*, *1895*, 55–63.

Ewing, R., Tian, G., & Lyons, T. (2018). Does compact development increase or reduce traffic congestion? *Cities*, *72*, 94–101.

Ewing, R., Tian, G., Spain, A., & Goates, J. P. (2014). Effects of light-rail transit on traffic in a travel corridor. *Journal of Public Transportation*, *17*(4), 93–113.

Ewing, R., Zhang, M., & Greenwald, M. (2011). School trips: Analysis of factors affecting mode choice in three metropolitan areas. In *School siting and healthy communities: Why where we invest in school facilities matters*. East Lansing, MI: Michigan State University Press.

Fargo-Moorhead Council of Governments. (2009). *Long range transportation plan*. Retrieved from www.fmmetrocog.org/index.php?option=com_docman&task=cat_view&gid=21&Itemid=3

Fargo-Moorhead Metropolitan Council of Governments. (2014). *2014 long range transportation plan*. Retrieved from http://www.fmmetrocog.org/new/assets/documents/LRTP/2014%20Long%20Range%20TransportationPlan%20-%20Metro%202040%20Approved%20071714.pdf

Farhan, B., & Murray, A. T. (2008). Siting park-and-ride facilities using a multi-objective spatial optimization model. *Computers & Operations Research*, *35*(2): 445–456.

Federal-Aid Highway Act of 1934, Ch. 586, §11, 73 Pub. L. No. 393, 48 Stat. 993.

Federal-Aid Highway Act of 1956, Ch. 462, §108, 84 Pub. L. No. 627, 70 Stat. 374.

Federal-Aid Highway Act of 1962, 87 Pub. L. No. 866, 76 Stat. 1145.

Federal-Aid Highway Act of 1973, 93 Pub. L. No. 87; 87 Stat. 250.

Federal Highway Administration (FHWA). (1976). *America's Highways 1776-1976*. Washington, DC: U.S. Department of Transportation.

Federal Highway Administration (FHWA). (1993). *Statewide Planning; Metropolitan Planning, 58 Fed. Reg. 58040, 58072 (*Oct. 28, 1993) (codified at 23 C.F.R. §450.316(a)(4)) (2006).

Federal Highway Administration (FHWA). (1995). *Highway statistics summary to 1995*. Washington, DC: Office of Highway Information Management.

Federal Highway Administration (FHWA). (2003). *Urban boundary and federal functional classification handbook*. Washington, DC: US Department of Transportation.

Ferguson, E. T., & Sanford, E. L. (1991). *Trip reduction ordinances: An overview*. Paper presented at the 70th Annual Meeting, Transportation Research Board. Washington, DC.

Ferreira, A. (2012). Accessibility is gold, mobility is not: A proposal for the improvement of Dutch transport-related cost: Benefit analysis. *Environment and Planning B: Planning and Design*, *39*(4), 683–697.

Filipova, A. A. (2006). *Facing the future: Georgia Regional Transportation Authority re-examined*. Presented at the Midwestern Policy Science Association 2006 Annual Meeting, Chicago, IL (pp. 1–32).

Finn, P. (1997). *VRE, facing 99 shortfall, is urged to think smaller*. Washington, DC: The Washington Post.

Flagstaff Metropolitan Planning Organization. (2011). *Regional transportation coordination plan*. Retrieved from www.flagstaff.az.gov/DocumentView.aspx?DID=6797

Florida Department of Transportation (FDOT). (2006). *Median handbook: Interim version*. Tallahassee: Florida Department of Transportation.

Frank, L. D. (2000). Land use and transportation interaction: Implications on public health and quality of life. *Journal of Planning Education and Research*, *20*(1), 6–22.

Frank, L. D., & Engelke, P. (2001). The built environment and human activity patterns: Exploring the impacts of urban form on public health. *Journal of Planning Literature*, *16*(2), 202–218.

Fresno Council of Governments. (2014). *Regional transportation plan and sustainable communities strategy*. Retrieved from https://www.fresnocog.org/project/regional-transportation-plan-rtp/

Garceau, T. J., Atkinson-Palombo, C., & Garrick, N. (2015). Peak car travel in the United States: Two-decade-long phenomenon at the state level. *Transportation Research Record: Journal of the Transportation Research Board*, (2531), 36–44, Washington, DC: Transportation Research Board.

Garreau, J. (1981). *The Nine Nations of North America*. Boston, MA: Houghton Mifflin.

Garrett, M., & Wachs, M. (1996). *Transportation planning on trial: The Clean Air Act and travel forecasting*. Newbury Park, CA: Sage Publications.

Gebel, K., Bauman, A. E., & Petticrew, M. (2007). The physical environment and physical activity: A critical appraisal of review articles. *American Journal of Preventive Medicine*, *32*(5), 361–369.

Geller, R. (2009). *Four types of cyclists*. Portland, OR: Portland Office of Transportation. Retrieved from www.portlandoregon.gov/transportation/article/237507

Genesee Transportation Council. (2011). *Long range transportation plan for the Genesee-Finger Lakes Region*. Retrieved from http://www.gtcmpo.org/LRTP

Georggi, N. L., Winters, P., Rai, S., & Zhou, L. (2007). Measuring the impacts of employer-based transportation demand management programs on an interstate corridor. *Journal of Public Transportation*, *10*(4), 51–78.

Geurs, K. T., & van Wee, B. (2004). Accessibility evaluation of land-use and transport strategies: Review and research directions. *Journal of Transport Geography*, *12*(2), 127–140.

Gibson, S. D., & Ambroz, J. (2008). *Integration of local agency pavement management system data to prioritize regional road project selection*. TRB 87th Annual Meeting Compendium of Papers DVD. Retrieved from http://pubsindex.trb.org/orderform.html

Giles, L. (1910). *Sun Tzu on the art of war: The oldest military treatise in the world*. Champaign, IL: Project Gutenberg.

Goldman, T., & Gorham, R. (2006, January). Sustainable urban transport: Four innovative directions. *Technology in Society*, *28*(1–2), 261–273.

Goodwin, P., Dargey, J., & Hanly, M. (2004). Elasticities of road traffic and fuel consumption with respect to price and income: A review. *Transport Reviews*, *24*(3), 275–292.

Goodwin, P., & Van Dender, K. (2013). "Peak car": Themes and issues. *Transport Reviews*, *33*(3), 243–254.

Gorham, R. (2009). Demystifying induced travel demand: Sustainable urban transport technical document No. 1. Bonn, Germany.

Government Accountability Office (GAO). (2005). *Highway congestion: Intelligent transportation systems' promise for managing congestion falls short, DOT could better facilitate their strategic use*. Washington, DC: United States Government Accountability Office.

Grace, J. B. (2006). *Structural equation modeling and natural systems*. United Kingdom: Cambridge University Press.

Graham, D. J., & Glaister, S. (2004). Road traffic demand elasticity estimates: A review. *Transport Reviews*, *24*(3), 261–274.

Grand Valley Metropolitan Council. (2011). *2035 long range transportation plan for the Grand Rapids metropolitan area*. Retrieved from www.gvmc.org/transportation/longrangeplan.shtml#2035LRTPU

Grand Valley Metropolitan Council. (2011). *2035 long range transportation plan*. Retrieved from https://static1.squarespace.com/static/59dce13bb1ffb65b4d405588/t/5a6f6a38e4966b1bdcca7155/1517251163515/2035_LRTP_Update_Final.pdf

Greater Bridgeport Regional Council. (2011). *Regional transportation plan for the Greater Bridgeport planning region: 2011-2040*. Retrieved from http://www.ctmetro.org/uploads/PDFs/Publications/Reports/Transportation/LRTP/GBRC-Greater-Bridgeport-CT-Transportation-Plan-2011-2040.pdf

Greater Buffalo-Niagra Regional Transportation Council. (2005). *2030 long range plan*. Retrieved from www.gbnrtc.org/planning/lrp/

Greater Buffalo Niagara Regional Transportation Council. (2014). *2040 metropolitan transportation plan update*. Retrieved from https://static1.squarespace.com/static/56ccbbfd3c44d8670dbd1d84/t/56f051e620c647792e8f757d/1458590301205/2040+Metropolitan+Transportation+Plan+Update+Final.pdf

Greenville-Pickens Area Transportation Study. (2013). *Long range transportation plan update*. Retrieved from http://www.gpats.org/_uploads/2011/12/2035LRTPUpdate_Nov2013_DRAFT_11.6.13.pdf

Greenwald, M. J. (2003). The road less traveled: New Urbanist inducements to travel mode substitution for nonwork trips. *Journal of Planning Education and Research, 23*(1), 39–57.

Greenwald, M. J. (2006). The relationship between land use and intrazonal trip making behaviors: Evidence and implications. *Transportation Research Part D: Transport and Environment, 11*(6), 432–446.

Halden, D. (2011, November). The use and abuse of accessibility measures in UK passenger transport planning. *Research in Transportation Business & Management, 2*, 12–19.

Halden, D. (2014). Shaping the future: Case studies in UK accessibility planning. *Transportation Research Procedia, 1*, 284–292.

Hall, P. (1998). *Cities in civilization*. London, UK: Fromm International.

Hamidi, S., & Ewing, R. (2014). A longitudinal study of changes in urban sprawl between 2000 and 2010 in the United States. *Landscape and Urban Planning, 128*, 72–82.

Hampton Roads Transportation Planning Organization. (2012). *2034 long-range transportation plan*. Retrieved from https://www.hrtpo.org/uploads/docs/Reports/Final_2034LRTP.pdf

Handy, S. L. (2008). Regional transportation planning in the US: An examination of changes in technical aspects of the planning process in response to changing goals. *Transport Policy, 15*(2), 113–126.

Handy, S. L., & Xing, Y. (2011). Factors correlated with bicycle commuting: A study in six small U.S. cities. *International Journal of Sustainable Transportation, 5*(2), 91–110.

Hansen, M. H. (1998). *Polis and city-state: An ancient concept and its modern equivalent*. Munksgaard, Denmark: Munksgaard.

Hansen, W. G. (1959, May). How accessibility shapes land use. *Journal of the American Institute of Planners, 25*(2), 73–76.

Hanson, S. (2004). The context of urban travel. In S. Hanson & G. Giuliano (Eds.), *The geography of urban transportation* (3rd ed., pp. 3–29). New York: Guilford Press.

Hartgen, D. T., & Fields, M. G. (2006). *Building roads to reduce traffic congestion in America's cities: How much and at what cost?* Los Angeles, CA: Reason Foundation.

Hatzopoulou, M., & Miller, E. J. (2009). Transport policy evaluation in metropolitan areas: The role of modelling in decision-making. *Transportation Research Part A, 43*, 323–338.

Haubrich, J. G., & Meyer, B. (2007). *Peak oil*. Cleveland, OH: Federal Reserve Bank of Cleveland.

Heath, G. W., Brownson, R. C., Kruger, J., Miles, R., Powell, K. E., Ramsey, L. T., & the Task Force on Community Preventive Services. (2006). The effectiveness of urban design and land use and transport policies and practices to increase physical activity: A systematic review. *Journal of Physical Activity and Health, 3*(1), 55–76.

Heggie, I. G., & Vickers, P. (1998). *Commercial management and financing of roads* (World Bank Technical Paper 409). Washington, DC: World Bank.

Henry, L., & Litman, T. A. (2006). *Evaluating new start transit program performance: Comparing rail and bus*. Victoria, BC: Victoria Transport Policy Institute.

Hess, D., Taylor, B., & Yoh, A. (2005). Light rail lite or cost-effective improvements to bus service? Evaluating costs of implementing bus rapid transit. *Transportation Research Record: Journal of the Transportation Research Board, 1927*, 22–30.

Hidalgo County Metropolitan Planning Organization. (2010). *2010-2035 metropolitan transportation plan*. Retrieved from http://oaktrust.library.tamu.edu/handle/1969.1/90600

Hillsborough County Metropolitan Planning Organization. (2009). *Hillsborough County 2035 long range transportation plan*. Retrieved from http://www.planhillsborough.org/wp-content/uploads/2012/08/2035-Plan-Cover-Contents-and-Introduction.pdf

HLB Decision Economics. (2001). *The economic and community benefits of transportation options for greater Cincinnati*. Cincinnati: Ohio-Kentucky-Indiana Metropolitan Planning Organization.

Holland, D. K. (1974). *A review of reports relating to the effects of fare and service changes in metropolitan public transit systems*. St. Louis, MO: Center for Urban Programs, St. Louis University.

Holmes, O. W., Jr. (2004). *Speeches by Oliver Wendell Holmes*. Whitefish, MT: Kessinger Publishing.

House Committee on Transportation and Infrastructure. (1991). *House of representatives report No. 102–171*. Pt. 1. Washington, DC: U.S. Government Publishing Office. Reprinted in 1991 U.S.C.C.A.N. 1526.

House Committee on Transportation and Infrastructure. (1998). *House of representatives conference report No. 105-550*. Washington, DC: U.S. Government Publishing Office. Reprinted in 1998 U.S.C.C.A.N. 70.

House of Commons. (2013). *Transport and accessibility to public services*. Environmental Audit Committee 3rd Report. London: Stationary Office Limited.

Houston-Galveston Area Council. (2007). *Bridging our communities 2035: The 2035 Houston-Galveston regional transportation plan*. Retrieved from www.h-gac.com/taq/plan/default.aspx

Houston-Galveston Area Council. (2015). *Bridging our communities: The Houston-Galveston regional transportation plan*. Retrieved from http://www.h-gac.com/taq/plan/2040/docs/2040-RTP-revised-April-2016.pdf

Howitt, A. M., & Moore, E. M. (1999). *Linking transportation and air quality planning: Implementation of the transportation conformity regulations in 15 nonattainment areas*. Cambridge, MA: Harvard University.

Humpel, N., Owen, N., & Leslie, E. (2002). Environmental factors associated with adults' participation in physical activity: A review. *American Journal of Preventative Medicine*, *22*(3), 188–199.

Indian Nation Council of Governments. (2005). *Destination 2030: Long range transportation plan*. Retrieved from https://www.pagnet.org/tabid/382/default.aspx

Indianapolis Metropolitan Planning Organization. (2009). *The Indianapolis 2030 regional transportation plan*. Retrieved from www.indympo.org/Plans/Regional/Pages/home.aspx

Indianapolis Metropolitan Planning Organization. (2011). *2035 long-range transportation plan*. Retrieved from https://d16db69sqbolil.cloudfront.net/mpo-website/downloads/LRTP/LRTP-2035/Central-Indiana-2035-LRTP.pdf

Intermodal Surface Transportation Efficiency Act of 1991, 102 Pub. L. No. 240, 105 Stat. 1914.

International Energy Agency (IEA). (2017). *World Energy Outlook 2017*. Paris: Organisation for Economic Co-operation and Development, OECD.

Iseki, H., & Taylor, B. D. (2009). Not all transfers are created equal: Towards a framework relating transfer connectivity to travel behaviour. *Transport Reviews*, *29*(6), 777–800.

Iseki, H., & Taylor, B. D. (2010). Style versus service? An analysis of user perceptions of transit stops and stations. *Journal of Public Transportation*, *13*(3), 2.

Jack Faucett Associates. (1999). *Granting air quality credit for land use measures: Policy options*. Washington, DC: U.S. Environmental Protection Agency.

Jacobs, J. (1961). *The death and life of great American cities*. New York: Vintage.

Jaroszynski, M., & Brown, J. (2014). *Do LRT planning decisions affect metropolitan transit performance? An examination of 8 US metropolitan areas with LRT transit backbones*. Transportation Research Record.

Johnston, R. A. (2004). The urban transportation planning process. In S. Hanson & G. Guiliano (Eds.), *The geography of urban transportation* (pp. 115–140). New York: Guilford Press.

Johnston, R. A. (2006). *Review of U.S. and European regional modeling studies of policies intended to reduce motorized travel, fuel use, and emissions*. Victoria Transport Policy Institute. Retrieved from www.vtpi.org/johnston.pdf

Johnston, R. A., & Ceerla, R. (1996). The effects of new high-occupancy vehicle lanes on travel and emissions. *Transportation Research Part A*, *30*(1), 35–50.

Johnston, R. A., Rodier, C. J., Choy, M., & Abraham, J. E. (2000). *Air quality impacts of regional land use policies*. Washington, DC: U.S. Environmental Protection Agency.

Kahn, E. B., Ramsey, L. T., Brownson, R. C., Heath, G. W., & Howze, E. H. (2002). The effectiveness of interventions to increase physical activity: A systematic review. *American Journal of Preventive Medicine*, *22*(4), 73–107.

Kahn, H. (1962). *Thinking about the unthinkable*. New York: Horizon Press.

Kain, J. F., & Liu, Z. (1996). *Econometric analysis of determinants of transit ridership: 1960–1990*. Prepared for the Volpe National Transport Systems Center. Cambridge, MA: US Department of Transportation.

Kalyvitis, S., & Vella, E. (2011). Public capital maintenance, decentralization, and US productivity growth. *Public Finance Review*, 39(6), 784–809.

Katz, B., & Bradley, J. (2013). *The metropolitan revolution: How cities and metros are fixing our broken politics and fragile economy*. Washington, DC: Brookings Institution Press.

Katz, B., Puentes, R., & Bernstein, S. (2003). *TEA-21 reauthorization: Getting transportation right for metropolitan America*. Washington, DC: Brookings Institution Press.

Kemmis, D. (2001). *This sovereign land: A new vision for governing the West*. Washington, DC: Island Press.

Kern Council of Governments. (2014). *2014 regional transportation plan/sustainable communities strategy*. Retrieved from https://www.arb.ca.gov/cc/sb375/kerncog_2014_rtp.pdf

Knoxville Regional Transportation Planning Organization. (2013). *Long range regional mobility plan 2040*. Retrieved from https://knoxtrans.org/plans/mobilityplan/archives/2013.pdf

Kotler, M. (2005). *Neighborhood government: The local foundations of political life*. Lanham, MD: Lexington Books.

Krahnstoever-Davison, K., & Lawson, C. T. (2006). Do attributes in the physical environment influence children's physical activity? A review of the literature. *International Journal of Behavioral Nutrition and Physical Activity*, *3*(19).

Krizek, K. J. (2003a). Residential relocation and changes in urban travel: Does neighborhood-scale urban form matter? *Journal of the American Planning Association*, *69*(3), 265–281.

Krizek, K. J. (2003b). Operationalizing neighborhood accessibility for land use-travel behavior research and regional modeling. *Journal of Planning Education and Research*, *22*(3), 270–287.

Krizek, K. J. (2003c). Neighborhood services, trip purpose, and tour-based travel. *Transportation*, *30*(4), 387–410.

Krizek, K. J., Barnes, G., Wilson, R., Johns, B., McGinnis, L., NuStats International, Forsyth, A., Handy, S. L., & Clifton, K. J. (2007). *Nonmotorized transportation pilot program evaluation study: Final report*. St. Paul, MN: Minnesota Department of Transportation, University of Minnesota.

Krizek, K. J., El-Geneidy, A., & Thompson, K. (2007). A detailed analysis of how an urban trail system affects cyclists' travel. *Transportation*, *34*, 611–624.

K.T. Analytics. (1995). *Parking management strategies: A handbook for implementation, for regional transportation authority*, Chicago, Illinois, in association with Barton-Aschman Associates and Eric Schreffler.

Kulash, D. J., & Urban Institute. (1974). *Parking taxes for congestion relief : a survey of related experience*. Washington, D.C.: Urban Institute.

Kuzmyak, J. R., Evans, J. E., et al. (2010). Traveler response to transportation system changes. In *Employer and institutional TDM strategies* (Transit Cooperative Research Program Report 91; Chapter 19). Washington, DC: Transportation Research Board. Retrieved from www.trb.org/Publications/Blurbs/163781.aspx

Lackawanna County Planning Commission, Luzerne County Planning Commission, & Pennsylvania Department of Transportation. (2011). *Lackawanna-Luzerne Regional Plan*. Retrieved from http://www.luzernecounty.org/county/departments_agencies/planning_commission/lackawanna-luzerne-regional-plan

Lancaster County. (2012). *Connections 2040: The comprehensive plan for Lancaster County, Pennsylvania*. Retrieved from https://lancastercountyplanning.org/133/Connections

Lane, M. B. (2005). Public participation in planning: An intellectual history. *Australian Geographer*, *36*(3), 283–299.

Lang, R. E. (2003). *Edgeless cities: Exploring the elusive metropolis*. Washington, DC: Brookings Institution Press.

Lau, L. (2012). *Best planning practices: Metropolitan transportation plans*. U.S. Department of Transportation/Volpe National Transportation Systems Center. Retrieved from www.planning.dot.gov/documents/BestPlanningPractices_MTP.pdf

Lebanon County Metropolitan Planning Organization. (2012). *Lebanon County 2013-2035 long range transportation plan*. Retrieved from http://www.lebcounty.org/depts/Planning/Documents/MPO_Documents/LEBCOMPOLRTPfor2017TIP(1).pdf

Lee County Metropolitan Planning Organization. (2010). *2035 long range transportation plan*. Retrieved from http://leempo.com/programs-products/long-range-transportation-plan/

Lee, C., & Moudon, A. V. (2004). Physical activity and environment research in the health field: Implications for urban and transportation planning practice and research. *Journal of Planning Literature, 19*, 147–181.

Lehigh Valley Planning Commission. (2010). *Lehigh Valley surface transportation plan 2011-2030*. Retrieved from http://www.lvpc.org/pdf/surfaceTransPlan2011-2030/surfaceTransPlan.pdf

Leinberger, C. B. (2009). *The option of urbanism: Investing in a New American Dream*. Washington, DC: Island Press.

Leinberger, C. B., & Lynch, P. (2014). *Foot traffic ahead: Ranking walkable urbanism in America's largest metros*. Washington, DC: Center for Real Estate and Urban Analysis.

Levine, J., Grengs, J., & Shen, Q. (2009). *Metropolitan accessibility and transportation sustainability: Comparative indicators for policy reform*. Retrieved from http://onlinepubs.trb.org/onlinepubs/conferences/2010/livability/Levine.pdf

Levine, J., Grengs, J., & Shen, Q. (2012). Does accessibility require density or speed? A comparison of fast versus close in getting where you want to go in U.S. metropolitan regions. *Journal of the American Planning Association, 78*(2), 157–172.

Lewis, P. G., & Sprague, M. (1997). *Federal transportation policy and the role of metropolitan planning organizations in California*. San Francisco, CA: Public Policy Institute of California.

Lindley, J. A. (1987). Urban freeway congestion: Quantification of the problem and effectiveness of potential solutions. *ITE Journal, 57*, 27–32.

Litman, T. A. (2007). *Transportation elasticities: How prices and other factors affect travel behavior*. Retrieved from www.vtpi.org/tdm/tdm11.htm

Litman, T. A. (2009a). *Smart congestion reductions: Reevaluating the role of highway expansion for improving urban transportation*. Victoria Transport Policy Institute. Retrieved January 14, 2010 from www.vtpi.org/documents/evaluation.php_

Litman, T. A. (2009b). *Transportation cost and benefit analysis: Techniques, estimates, and implications* (2nd ed.). Victoria Transport Policy Institute. Retrieved February 9, 2010 from www.vtpi.org/tca/tca01.pdf

Litman, T. A. (2011). *Sustainability and livability: Summary of definitions, goals, objectives and performance Indicators*. Victoria, BC: Victoria Transport Policy Institute.

Litman, T. A. (2013). The new transportation planning paradigm. *Institute of Transportation Engineers. ITE Journal, 83*(6), 20.

Litman, T. A. (2015). *Understanding smart growth savings, Victoria Transport Policy Institute*. Retrieved from www.vtpi.org/sg_

Litman, T. A., & Fitzroy, S. (2005). *Safe travels: Evaluating mobility management traffic safety impacts*. Victoria, BC: Victoria Transport Policy Institute.

Louisville Metro Government. (2016). *2035 transportation plan*. Retrieved from https://louisvilleky.gov/sites/default/files/advanced_planning/movelouisville_april2016draft.pdf

Lowe, K. (2014). *Regional decision-making and competitive funding: Metropolitan planning organizations and the transportation investments generating economic recovery program*. National Technical Information Service, University of New Orleans, Southwest Region University Transportation Center, Research and Innovative Technology Administration. Retrieved from http://swutc.tamu.edu/pub...reports/600451-00106-1.pdf

Lusk, A. C., Furth, P. G., Morency, P., Miranda-Moreno, L. F., Willett, W. C., & Dennerlein, J. T. (2011). *Risk of injury for bicycling on cycle tracks versus in the street*. Brief Report. Retrieved from http://injuryprevention.bmj.com/content/early/2011/02/02/ip.2010.028696.full Accessed 24 July 2013.

Macbeth, A. G. (1999). Bicycle lanes in Toronto. *ITE Journal*, *69*, 4.

Madison Area Transportation Planning Board. (2010). *Regional transportation plan for 2030*. Retrieved from www.tri-co.org/trp.htm#

Madison Area Transportation Planning Board. (2012). *Regional transportation plan update*. Retrieved from http://www.madisonareampo.org/planning/documents/RTPFINAL_Web.pdf

Madsen, T., Davis, B., & Baxandall, P. (2010). *Road work ahead: Holding government accountable for fixing America's crumbling roads and bridges*. Boston: U.S. PIRG Education Fund. Retrieved from www.uspirg.org/reports/usp/road-work-ahead

Margerum, R. (2010). *Regional transportation and land use decision making: A multistate analysis*. Oregon Department of Transportation, University of Oregon, Eugene. Retrieved from http://otrec.us/project/340

Maricopa Association of Governments. (2003). *Regional transportation plan*. Retrieved from www.azmag.gov/Projects/Project.asp?CMSID2=1126&MID=Transportation

Maricopa Association of Governments. (2014). *2035 regional transportation plan*. Retrieved from http://www.azmag.gov/Portals/0/Documents/RTP_2014-01-30_Final-2035-Regional-Transportation-Plan-(RTP)-Executive-Summary.pdf?ver=2017-04-06-111659-413

Martens, M. J., & Griethuysen, S. V. (1999). *The ABC location policy in the Netherlands "the right business at the right place"*. The Hague, Netherlands: TNO Inro.

Mazziotti, D. F., Hemphill, M., Churchill, L., Hamilton, J., & Gies, M. (1977). *Energy conservation choices for the city of Portland, Oregon*. Washington, DC: U.S. Department of Energy.

McCarthy, J. E. (2004). *Transportation conformity under the Clean Air Act: In need of reform?* Washington, DC: Congressional Research Service. Retrieved from http://ncseonline.org/NLE/CRSreports/04apr/RL32106.pdf

McCormack, G., Giles-Corti, B., Lange, A., Smith, T., Martin, K., & Pikora, T. J. (2004). An update of recent evidence of the relationship between objective and self-report measures of the physical environment and physical activity behaviours. *Journal of Science and Medicine in Sport*, *7*(1), 81–92.

McDowell, B. D. (1984). The metropolitan planning organization role in the 1980s. *Journal of Advanced Transportation*, *18*, 125–133.

McKinsey and Company. (2008). *IT3 scenario results and implications: Briefing to the general assembly*. Atlanta: State of Georgia.

McMullen, B. S., & Eckstein, N. (2012). Relationship between vehicle miles traveled and economic activity. *Transportation Research Record: Journal of the Transportation Research Board*, (2297), 21–28. Washington, DC: Transportation Research Board.

Mecklenburg-Union Metropolitan Planning Organization. (2010). *2035 long-range transportation plan*. Retrieved from www.mumpo.org/2035_LRTP.htm

Metroplan Orlando. (2016). *2040 long range transportation plan*. Retrieved from https://metroplanorlando.org/wp-content/uploads/2040-lrtp-plan-overview-1.pdf

Metroplan (Arkansas). (2010). *Metro 2030 long range transportation plan*. Retrieved from www.metroplan.org/index.php?fuseaction=p0007.&mod=44

Metropolitan Area Planning Agency. (2015). *2035 long-range transportation plan*. Retrieved from http://mapacog.org/wp-content/uploads/2015/10/LRTP_2035_Amendment2.pdf
Metropolitan Transportation Commission. (2009). *Change in motion: Transportation 2035 plan for the San Francisco Bay Area*. Retrieved from https://mtc.ca.gov/our-work/plans-projects/plan-bay-area-2040/transportation-2035
Metropolitan Transportation Commission. Change in Motion. Retrieved from https://mtc.ca.gov/sites/default/files/4_Investments-final.pdf
Metropolitan Washington Council of Governments. (2002). *State of the commute 2001: Survey results from the Washington metropolitan region*. Washington, DC: MWCOG.
Miami-Dade Metropolitan Planning Organization. (2009). *Miami-Dade 2035 long range transportation plan*. Retrieved from http://www.miamidadetpo.org/long-range-transportation-plan.asp
Miami Valley Regional Planning Council. (2008). *2030 long range transportation plan*. Retrieved from www.mvrpc.org/transportation/long-range
Miami Valley Regional Planning Commission. (2012). *2040 long range transportation plan*. Retrieved from https://www.mvrpc.org/transportation/long-range-planning-lrtp
Michael Baker Corporation, Crain & Associates, LKC Consulting Services, & Howard/Stein-Hudson. (1997). *The potential of public transit as a transportation control measure: Case studies and innovations*. Draft. Pittsburgh, PA: Michael Baker Corporation.
Mid-America Regional Council (2005). *Transportation outlook 2030 update*. Retrieved from http://www.marc.org/outlook2030/
Mid-America Regional Council. (2010). *Transportation outlook 2040*. Retrieved from http://www.to2040.org/
Mid-America Regional Council. (2015). *Transportation outlook 2040*. Retrieved from www.to2040.org/assets/2015_plan/1.0_ExecSumm_adopt_final.pdf
Mid-Ohio Regional Planning Commission. (2008). *CapitalWays*. Retrieved from www.morpc.org/transportation/mtp/mtp_main.asp
Mid-Ohio Regional Planning Commission. (2013). *2012-2035 Metropolitan transportation plan*. Retrieved from http://www.morpc.org/program-service/metropolitan-transportation-plan/
Mid-Region Council of Governments of New Mexico. (2007). *Metropolitan transportation plan*. Retrieved from www.mrcog-nm.gov/transportation-mainmenu-67/metro-planning-mainmenu-188/long-range-mtp-mainmenu-189/347-2030-mtp
Mid-Region Council of Governments of New Mexico. (2015). *Metropolitan transportation plan*. Retrieved from www.mrcog-nm.gov/images/stories/pdf/transportation/2040_MTP/2040-mtp-chapter-3.pdf
Mid-Region Metropolitan Planning Organization. (2011). *2035 metropolitan transportation plan*. Retrieved from https://www.mrcog-nm.gov/images/stories/pdf/transportation/2035_mtp/Final_Approved/CompleteMTPLocalFedApproved_May_2014.pdf
Miller, H. J. (2005). Place-based versus people-based accessibility. In D. Levinson & K. J. Krizek (Eds.), *Access to destinations* (pp. 63–89). Oxford, UK: Elsevier.
Minneapolis-St. Paul Metropolitan Council. (2010). *2030 transportation policy plan: Metropolitan Kansas City's long-range transportation plan*. Retrieved from

https://metrocouncil.org/Transportation/Publications-And-Resources/2030-Transportation-Policy-Plan-Chapter-6-Highways.aspx

Mokhtarian, P. L., Salomon, I., & Redmond, L. S. (2001). Understanding the demand for travel: It's not purely "derived". *Innovation*, *14*(4), 355–380.

Montgomery County, Maryland. (1989). *Comprehensive growth policy study*. Silver Spring, MD: Maryland-National Capital Park & Planning Commission.

Morris, A. J. E. (1996). *History of urban form: Before the industrial revolution* (3rd ed.). Upper Saddle River, NJ: Prentice Hall.

Moudon, A., Lee, C. Cheadle, A.D., Collier, C.W., Johnson, D., Schmid, T. L., & Weather, R. D. (2005). Cycling and the built environment, a US perspective. *Transportation Research Part D: Transport and Environment*, *10*, 245–261. 10.1016/j.trd.2005.04.001.

Mountainland Association of Governments. (2011). *2040 metropolitan transportation plan*. Retrieved from https://www.mountainland.org/transplan40-document-category

Mumford, L. (1938). *The culture of cities*. New York: Harcourt Brace Jovanovich.

Nashville Area Metropolitan Planning Organization. (2010). *Nashville area regional transportation plan*. Retrieved from http://www.nashvillempo.org/docs/lrtp/2035rtp/Docs/2035_Doc/2035Plan_Complete.pdf

National Archives and Records Administration (NARA). (2014, June 2). Statewide and nonmetropolitan transportation planning; metropolitan transportation planning. *Federal Register*, *79, 105*(4).

National Association of City Transportation Officials (NACTO). (2012). *NACTO urban bikeway design guide*. Washington, DC: National Association of City Transportation Officials. Retrieved from http://nacto.org/cities-for-cycling/design-guide/

National Capital Region Transportation Planning Board. (2014). *Financially constrained long-range transportation plan for the National Capital Region*. Retrieved from https://www.mwcog.org/transportation/plans/clrp/

National Interregional Highway Committee. (1944). *House of representatives document no. 78–379*. Washington, DC: Government Printing Office.

National Research Council. (2005). *Does the built environment influence physical activity? Examining the evidence*. (Special Report 282). Washington, DC: Transportation Research Board and Institute of Medicine Committee on Physical Activity, Health, Transportation, and Land Use.

Nelson, A. C. (2000). New kid in town: The Georgia regional transportation authority and its role in managing growth in metropolitan Georgia. *Wake Forest Law Review*, *35*, 625.

Nelson, A. C., & Allen, D. (1997). If you build them, commuters will use them. *Transportation Research Record*, *1578*, 79–83.

Nelson, A. C., & Lang, R. E. (2011). *Megapolitan America: A new vision for understanding America's metropolitan geography*. Chicago, IL: American Planning Association Press.

Newman, P., & Kenworthy, J. (1999). *Sustainability and cities: Overcoming automobile dependence*. Washington, DC: Island Press.

New State Ice Co. v. Liebmann, 285 U.S. 262, 280 (1932).

New York Metropolitan Transportation Council. (2010). *2010–2035 NYMTC region transportation plan: Executive summary: A shared vision for a shared future*. Retrieved from www.nymtc.org/rtp/documents/CHAPTER/0_NYMTC_RTP_ExecSummary.pdf

New York Metropolitan Transportation Council. (2013). *Regional transportation plan: A shared vision for a sustainable region*. Retrieved from https://www.nymtc.org/Portals/0/Pdf/RTP/Plan%202040%20Main%20Document.pdf

Nichols, M. D. (2009). California's climate change program: Lessons for the nation. *UCLA Journal of Environmental Law and Policy*, *27*, 185.

Niedzielski, M. A., & Boschmann, E. (2014). Travel time and distance as relative accessibility in the journey to work. *Annals of the Association of American Geographers*, *104*(6), 1156–1182.

Niewiem, S. (2005). *The contractual relationship between clients and management consultants: A transaction cost economic analysis*. New York: DUV.

Noland, R. B. (2001). Relationships between highway capacity and induced vehicle travel. *Transportation Research Part A: Policy and Practice*, *35*(1), 47–72.

North Central Texas Council of Governments. (2009). *Mobility 2030: The metropolitan transportation plan for the Dallas-Fort Worth Area, 2009 Amendment*. Retrieved from www.nctcog.org/trans/mtp/2030/2009Amendment.asp

North Central Texas Council of Governments. (2013). *Mobility 2035: The metropolitan transportation plan for North Central Texas*. Retrieved from http://www.nctcog.org/trans/mtp/2035/m2035document.pdf

North Florida Transportation Planning Organization. (2010). *Envision 2035: Long range transportation plan*. Retrieved from http://northfloridatpo.com/global/resources/

North Jersey Transportation Planning Authority. (2013). *Plan 2040*. Retrieved from http://www.njtpa.org/archive/planning-archive/plan-update-to-2040/plan2040final

Northern Virginia Transportation Authority. (2006). *Trans action 2030: Northern Virginia's long range transportation plan*. Retrieved from www.thenovaauthority.org/transaction2030/ReportsandMaps/Trans2030-April-2006-Draft-Plan.pdf

Northern Virginia Transportation Authority. (2012). *TransAction 2040: Transportation for today and tomorrow*. Retrieved from http://thenovaauthority.org/planning/long-range-transportation/transaction-2040/

Oahu Metropolitan Planning Organization. (2006). *Oahu regional transportation plan 2030*. Retrieved from www.oahumpo.org/programs/ortp.html

Oahu Metropolitan Planning Organization. (2011). *Oahu regional transportation plan 2035*. Retrieved from http://www.oahumpo.org/?wpfb_dl=380

O'Connor, A. (1999). Swimming against the tide: A brief history of federal policy in poor communities. In J. DeFilippis & S. Saegert (Eds.), *The community development reader* (pp. 77–137). New York and London: Routledge.

Ohio-Kentucky-Indiana Council of Governments. (2008). *2030 regional transportation plan update*. Retrieved from www.oki.org/transportation/2030plan.html

Ohio-Kentucky-Indiana Council of Governments. (2012). *OKI 2040 regional transportation plan*. Retrieved from http://www.oki.org/departments/transportation/pdf/2040plan/finalchapters/fulldocument.pdf

Olson, B. K. (2000). The transportation equity act for the 21st century: The failure of metropolitan planning organizations to reform federal transportation policy in metropolitan areas. *Transportation Law Journal*, *28*, 147–170.

Orange County Transportation Authority. (2006). *New directions: Charting the course for Orange County's future transportation system*. Retrieved from www.octa.net/uploadedfiles/Files/pdf/lrtp06.pdf

Orange County Transportation Authority. (2014). *Outlook 2035: Because mobility matters*. Retrieved from http://www.octa.net/LRTP/?terms=Long%20Range%20Transportation%20Plan

Owen, N., Humpel, N., Leslie, E., Bauman, A., & Sallis, J. F. (2004). Understanding environmental influences on walking: Review and research agenda. *American Journal of Preventive Medicine*, *27*(1), 67–76.

Paaswell, R. E. (2013). *Infrastructure investment decision making: Emerging roles of planning and sustainability urban public transportation systems*. American Society of Civil Engineers. Third International Conference on Urban Public Transportation Systems, Paris (pp. 395–404). http://dx.doi.org/10.1061/9780784413210.035

Páez, A., Scott, D. M., & Morency, C. (2012, November). Measuring accessibility: Positive and normative implementations of various accessibility indicators. *Journal of Transport Geography*, *25*, 141–153.

Palm Beach Metropolitan Planning Organization. (2014). *Directions 2040: Long range transportation plan*. Retrieved from http://www.palmbeachtpa.org/LRTP

Parker, K. M., Rice, J., Gustat, J., Ruley, J., Spriggs, A., & Johnson, C. (2013). Effect of bike lane infrastructure improvements on ridership in one New Orleans neighborhood. *Annals of Behavioral Medicine*, *45*, S101–S107.

Parsons Brinckerhoff Quade Douglas. (1996). *Transit and urban form: Commuter and light rail transit corridors: The land use connection* (Transit Cooperative Research Program Report 16). Washington, DC: Transportation Research Board.

Pas, E. (1995). The urban transportation planning process. In S. Hanson (Ed.), *The geography of urban transportation* (pp. 53–77). New York: Guilford.

Peat Marwick Main & Co. (1989). *Status of traffic mitigation ordinances*. Washington, DC: Urban Mass Transportation Administration.

Peskin, R. L., & Schofer, J. L. (1977). *The impacts of urban transportation and land use policies on transportation energy consumption* (Report No. DOT-05–50118). Washington, DC: Department of Transportation.

Pikes Peak Area Council of Governments. (2008). *Moving forward 2035 regional transportation plan*. Retrieved from www.movingforwardplan.org/

Pikes Peak Area Council of Governments. (2012). *Moving forward update: 2035 regional transportation plan*. Retrieved from http://www.ppacg.org/transportation/long-range-transportation-plan/

Pima Association of Governments. (2010a). *2040 regional transportation plan*. Retrieved from www.pagnet.org/

Pima Association of Governments. (2010b). *Congestion management process: Final report*. Retrieved from http://www.pagnet.org/documents/Transportation/TranspoPlanning/PAG-CMP-FinalReportwithAppendix.pdf

Pioneer Valley Metropolitan Planning Organization. (2012). *2012 regional transportation plan*. Retrieved from http://www.pvpc.org/sites/default/files/2012%20Final%20RTP_9_21_11_web.pdf

Pooler, J. A. (1995). The use of spatial separation in the measurement of transportation accessibility. *Transportation Research Part A: Policy and Practice*, *29*(6), 421–427.

Porter, M. (1985). *Competitive advantage: Creating and sustaining superior performance*. London: Free Press.

Portland Bureau of Transportation. (2012). *Bicycle count report 2012*. Online Report. Retrieved from www.portlandoregon.gov/transportation/article/448401

Portland Metro Regional Government. (2004). *Creating livable streets: Street design guidelines for 2040*. Retrieved from www.oregonmetro.gov/index.cfm/go/by.web/id=235

Portland Metro Regional Government. (2010). *Destination tomorrow: 2010-2035 regional long-range transportation plan.* Retrieved from http://www.pactsplan.org/long-range-transportation-planning/the-2010-destination-tomorrow-plan/

Portland Metro Regional Government. (2014). *2014 regional transportation plan.* Retrieved from https://www.oregonmetro.gov/sites/default/files/2015/05/29/RTP-2014-final.PDF

Poughkeepsie-Dutchess County Transportation Council. (2012). *Moving Dutchess: The 2040 metropolitan transportation plan for Dutchess County.* Retrieved from http://www.dutchessny.gov/CountyGov/Departments/TransportationCouncil/24004.htm

Pratt, R. H., & Bevis, H. W. (1971). *An initial Chicago north suburban transit improvement program 1971–1975* (Urban Mass Transit Administration). Washington, DC: U.S Department of Transportation.

Pratt, R. H., et al. (2004). *Traveler response to transportation system changes: Bus routing and coverage* (Transit Cooperative Research Program TCRP Report 95; Chapter 10). Washington, DC: Transportation Research Board.

Pratt, R. H., et al. (2012). *Traveler response to transportation system changes, Chapter 16: Pedestrian and bicycle facilities* (Transit Cooperative Research Program Report 91). Washington, DC: Transportation Research Board.

Pratt, R. M., Pederson, N. J., & Mather, J. J. (1977). *Traveler response to transportation system changes: A handbook for transportation planners.* Washington, D.C.: Urban Mass Transportation Administration.

Proffitt, D., Bartholomew, K., Ewing, R., & Miller, H. (2015, January). *Accessibility planning in American metropolitan areas: Are we there yet?* Presented at the Transportation Research Board Annual Meeting, Washington, DC.

Proffitt, D., Bartholomew, K., Ewing, R., & Miller, H. J. (2017). Accessibility planning in American metropolitan areas: Are we there yet?. *Urban Studies*, 0042098017710122.

Pucher, J., & Buehler, R. (2009). Integrating bicycling and public transport in North America. *Journal of Public Transportation*, *12*(3), 101–126.

Pucher, J., & Buehler, R. (2012). City cycling. Cambridge: MIT Press.

Pucher, J., & Dijkstra, L. (2000). Making walking and cycling safer: Lessons from Europe. *Transportation Quarterly*, *54*(3), 25–50.

Pucher, J., & Dijkstra, L. (2003). Promoting safe walking and cycling to improve public health: Lessons from The Netherlands and Germany. *American Journal of Public Health*, *93*(9), 1509–1516.

Puget Sound Regional Council. (2007). *Destination 2030 update: Metropolitan transportation plan for the Central Puget Sound Region.*

Puget Sound Regional Council. (2010). *Transportation 2040: Toward a sustainable transportation system.* Retrieved from www.psrc.org/sites/default/files/t2040finalplan.pdf

Queensland Transport. (2007). *Bicycle mode share to CBD and CBD fringe 1986 to 2006* (Transport Research and Analysis Centre). Retrieved from www.bv.com.au/file/Brisbane_R2W_1986to2006.pdf

Ramos, L. E., Brown, J. R., & Nixon, H. (2015). The transit performance of modern-era streetcars: A consideration of five U.S. cities. *Transportation Research Record*, *2534*, 57–67.

Rao, P. K. (2006). *The economics of transaction costs: Theory, methods and applications.* New York: Palgrave Macmillan.

Raudenbush, S. W., & Byrk, A. S. (2002). *Hierarchical linear models: Applications and data analysis methods* (2nd ed.). Thousand Oaks, CA: Sage Publications.

Recker, W. W., Chen, C., & McNally, M. G. (2001). Measuring the impact of efficient household travel decisions on potential travel time savings and accessibility gains. *Transportation Research Part A: Policy and Practice*, *5*(4), 339–369

Regional Planning Commission of Greater Birmingham. (2012). *2040 regional transportation plan*. Retrieved from http://www.rpcgb.org/wp-content/uploads/2016/09/2040_RTP_BhamMPO_Jan14-2015_download.pdf

Regional Planning Commission. (2010). *Metropolitan transportation plan: New Orleans urbanized area*. Retrieved from http://www.norpc.org/metropolitan_transportation_plan.html

Regional Transportation Commission of Southern Nevada. (2012). *Regional transportation plan: 2013-2035*. Retrieved from http://www.rtcsnv.com/wp-content/uploads/2012/10/Final_RTP-2013-35-Redetermination-0214131.pdf

Regional Transportation Commission of Washoe County. (2008). *Regional transportation plan: (2008–2030), regional transportation plan, (2031–2040) Illustrative facilities plan*. Retrieved from www.rtcwashoe.com/planning-7

Regional Transportation Commission of Washoe County, Nevada. (2013). *2035 regional transportation plan*. Retrieved from http://www.rtcsnv.com/wp-content/uploads/2012/10/Final_RTP-2013-35-Redetermination-0214131.pdf

Rhode Island Statewide Planning Program. (2012). *Transportation 2035: Long range transportation plan*. Retrieved from http://www.planning.ri.gov/documents/trans/LRTP%202035%20-%20Final.pdf

Richmond Area Metropolitan Planning Organization. (2012). *Plan 2035: The long-range transportation plan for the Richmond area*. Retrieved from http://www.richmondregional.org/TPO/LRTP/plan2035/plan2035_Chapters.htm

Research and Innovative Technology Administration (RITA). (2007). *Deployment statistics: Freeway management*. Retrieved from www.itsdeployment.its.dot.gov/SurveyOutline1.asp?SID=fm

Richmond Regional Planning District Commission. (2008). *2031 long range transportation plan*. Retrieved from www.richmondregional.org/Urban_Transp-MPO/MPO_Div_Cats/toc.htm

Rusk, D. (1993). *Cities without suburbs*. Washington, DC: Woodrow Wilson Center Press.

Sacramento Area Council of Governments. (undated). *I-PLACE3S: What is I-PLACE3S?* Retrieved from www.sacog.org/services/I-PLACE3S/

Sacramento Area Council of Governments. (2007). *Sacramento region blueprint transportation land use study: Special report*. Retrieved from www.sacregionblueprint.org/implementation/

Sacramento Area Council of Governments. (2008). *Sacramento region metropolitan transportation plan for 2035*. Retrieved from www.sacog.org/mtp/2035/final-mtp/

Sacramento Area Council of Governments. (2010). *Welcome to PLACE3S: The I-PLACE3S user guide*. Retrieved from www.sacregionblueprint.org/technology/

Sacramento Area Council of Governments. (2012). *Metropolitan transportation plan/sustainable communities strategy*. Retrieved from https://www.sacog.org/regional-plans

Saelens, B. E., & Handy, S. L. (2008). Built environment correlates of walking: A review. *Medicine & Science in Sports & Exercise*, *40*, 7.

Saelens, B. E., Sallis, J. F., & Frank, L. D. (2003). Environmental correlates of walking and cycling: Findings from the transportation, urban design, and planning literatures. *Annals of Behavioral Medicine*, *25*(2), 80–91.

Safe, Accountable, Flexible, Efficient Transportation Equity Act: A Legacy for Users, Pub. L. No. 109–59, 119 Stat. 1144 (2005).

Sale, K. (1980). *Human scale*. New York: Coward, McCann & Geoghegan.

San Antonio-Bexar Council of Governments. (2004). *Mobility 2030*. Retrieved from www.sametroplan.org/Plans/MTP/MTP2030.pdf

San Diego Association of Governments. (2007). *2030 San Diego regional transportation plan: Pathways for the future*. Retrieved from www.sandag.org/programs/transportation/comprehensive_transportation_projects/2030rtp/2007rtp_final.pdf

San Diego Association of Governments. (2011). *Our region, our future: 2050 regional transportation plan*. Retrieved from http://www.sandag.org/uploads/2050RTP/F2050rtp_all.pdf

San Francisco Municipal Transportation Authority. (2011). *2011 bicycle count report*. Retrieved from www.sfbike.org/download/bike_count_2011/2011BicycleCountReportsml_002.pdf

San Joaquin Council of Governments. (2014). *Regional transportation plan and sustainable communities strategy*. Retrieved from http://www.sjcog.org/DocumentCenter/View/489

Santa Barbara Association of Governments. (2008). *Vision 2030: Planning Santa Barbara County's transportation future*. Retrieved from www.sbcag.org/PDFs/publications/RTP_FINAL.pdf

Santa Barbara County Association of Governments. (2013). *2040 regional transportation plan and sustainable communities strategy*. Retrieved from http://www.sbcag.org/2013-rtp.html

Sarasota/Manatee Metropolitan Planning Organization. (2010). *2035 long range transportation plan*. Retrieved from https://www.mympo.org/documents

Schaub, C. (2010). *Prioritizing projects for the metropolitan transportation plan: The Lexington experience*. Tools of the Trade: 12th National Conference on Transportation Planning for Small-and Medium-Sized Communities. Retrieved from http://trid.trb.org/view/1251878

Schneider, J. B., Deffebach, C., & Cushman, K. (1984). Timed-transfer/transit center concept as applied in Tacoma/Pierce County, Washington. *Transportation Quarterly*, *38*(3), 393–402.

Schrank, D., Eisele, B., & Lomax, T. (2012). *TTI's 2012 urban mobility report*. United States: Texas A&M Transportation Institute.

Schrank, D., Eisele, B., Lomax, T., & Bak, J. (2015). *2015 urban mobility Scorecard*. College Station, TX: Texas Transportation Institute.

Schrank, D., & Lomax, T. (2009). *The 2009 urban mobility report*. Texas Transportation Institute. Retrieved from http://mobility.tamu.edu

Schreffler, E. N., Costa, T., & Moyer, C. B. (1996). Evaluating travel and air quality cost-effectiveness of transportation demand management projects. In *Transportation Research Record 1520* (pp. 11–18). Washington, D.C.: TRB, National Research Council.

Schwartz, P. (1991). *The art of the long view*. New York: Double Day.

Seltzer, E., & Carbonell, A. (2011). Planning regions. In E. Seltzer & A. Carbonell (Eds.), *Regional planning in America: Practice and prospect*. Cambridge, MA: Lincoln Institute of Land Policy.

Senate Committee on Environment and Public Works. (1991). Senate Report No. 102-71 (June 4, 1991), Part A, Section 113. Washington, DC: U.S. Government Publishing Office.

Shoup, D. (1997). *Evaluating the effects of parking cash out: Eight case studies: Final report*. Sacramento, CA: California Air Resources Board Research Division.

Silva, C. (2013, April). Structural accessibility for mobility management. *Progress in Planning*, *81*, 1–49.

Simpson, B. (2002). Issues in integrating land use and transport policy: Evidence from South-West Birmingham. *Geography*, *87*(4), 355–366.

Sjoberg, G. (1965, September). The origin and evolution of cities. *Scientific American*, 54–57.

Small, K. A., & Van Dender, K. (2007). Fuel efficiency and motor vehicle travel: The declining rebound effect. *Economic Journal*, *28*(1), 25–51.

Smart Growth America. (2014). *Repair priorities 2014: Transportation spending strategies to save taxpayer dollars and improve roads*. Washington, DC: Smart Growth America.

Solof, M. (1997). *History of metropolitan planning organizations*. Newark, NJ: North Jersey Transportation Planning Authority. Retrieved from www.njtpa.org/public_affairs/mpo_history/MPOhistory1998.pdf

South Central Regional Council of Governments. (2011). *South Central regional long range transportation plan*. Retrieved from http://scrcog.org/wp-content/uploads/reports/LRTP_April272011approved.pdf

Southeastern Wisconsin Regional Planning Commission. (2006). *A regional transportation system plan for Southeastern Wisconsin: 2035*. Retrieved from http://www.sewrpc.org/SEWRPC/Transportation/2035RegionalTransportationPlan.htm

Southern California Association of Governments. (1988). *Evaluation report: Telecommuting pilot project*. Los Angeles, CA: SCAG.

Southern California Association of Governments. (2008). *2008 regional transportation plan: Making the connections*. Retrieved from www.scag.ca.gov/rtp2008/final.htm

Southern California Association of Governments. (2011). *Regional transportation plan: 2012-2035*. Retrieved from http://rtpscs.scag.ca.gov/Documents/2012/final/f2012RTPSCS.pdf

Southern California Association of Governments. (2012). *2012–2035 regional transportation plan*. Retrieved from http://rtpscs.scag.ca.gov/Documents/2012/final/f2012RTPSCS.pdf

Southwestern Planning Commission. (2006). *2035 transportation and development plan for Southern Pennsylvania*. Retrieved from www.spcregion.org/pub_lrp.shtml

Southwestern Pennsylvania Commission. (2011). *2040 transportation and development plan for Southwestern Pennsylvania*. Retrieved from http://www.spcregion.org/trans_lrp.shtml

Sperber, D. (1998). *The city in Roman Palestine*. Oxford, UK: Oxford University Press.

Spokane Regional Transportation Council. (2007). *Metropolitan transportation plan*. Retrieved from www.srtc.org/08%20MTP%20Update.html

Spokane Regional Transportation Council. (2013). *Horizon 2040*. Retrieved from https://www.srtc.org/horizon-2040/

Springfield-Sangamon County Regional Planning Commission. (2010). *2035 long range transportation plan*. Retrieved from http://co.sangamon.il.us/departments/m-r/regional-planning-commission/program-areas/strategic-comprehensive-planning

Standing Advisory Committee on Trunk Road Assessment (SACTRA). (1994). *Trunk roads and the generation of traffic*. London, UK: Department of Transport.

Stanley, K., Gleason, E., & Kyte, M. (1982). *Planning, implementation and evaluation of a time-transfer system in Portland, Oregon's suburban Westside*. 1982 Annual Meeting of the Transportation Research Board, Washington, DC.

Steiner, R. L., Bond, A., Miller, D., & Shad, P. (2004). Future directions for multimodal areawide level of service handbook research and development. In *Florida Department of Transportation Office of Systems Planning (BC-354-78)*.

Stern, H. (2008). A necessary collision: Climate change, land use, and the limits of A.B. 32. *Ecology Law Quarterly*, *35*, 611–637.

Stevens, M. R. (2017). Does compact development make people drive less?. *Journal of the American Planning Association*, *83*(1), 7–18.

Stoker, P., & Ewing, R. (2014). Job–worker balance and income match in the United States. *Housing Policy Debate*, 24(2), 485–497.

Surface Transportation Policy Project. (2000). *Changing direction: Federal transportation*. Washington, DC: Surface Transportation Policy Project.

Syracuse Metropolitan Transportation Council. (2007). *Long range transportation plan*. Retrieved from www.smtcmpo.org/lrtp.asp

Syracuse Metropolitan Transportation Council. (2015). *2050 long range transportation plan*. Retrieved from http://www.smtcmpo.org/LRTP2050/LRTP2050.asp

Tang, L., & Thakuriah, P. V. (2012). Ridership effects of real-time bus information system: A case study in the City of Chicago. *Transportation Research Part C: Emerging Technologies*, *22*:146–161.

Taylor, B. D., & Fink, C. N. Y. (2003). *The factors influencing transit ridership: A review and analysis of the ridership literature*. UCLA Department of Urban Planning Working Paper. Los Angeles, CA: UCLA.

Taylor, B. D., & Fink, C. N. Y. (2013). Explaining transit ridership: What has the evidence shown? *Transportation Letters*, *5*(1), 15–26.

Taylor, B. D., Miller, D., Iseki, H., & Fink, C. (2009). Nature and/or nurture? Analyzing the determinants of transit ridership across US urbanized areas. *Transportation Research Part A*, *43*, 60–77.

Toledo Metropolitan Area Council of Governments. (2011). *On the move: 2035 transportation plan*. Retrieved from http://tmacog.org/Transportation/2035/2035%20Plan%20Update%202011/On_The_Move_2007_2035_Update_2011.pdf

Transportation Equity Act for the 21st Century, 105 Pub. L. No. 178, 112 Stat. 107 (1998).

Transportation Research Board (TRB). (2009). *Driving and the built environment: The effects of compact development on motorized travel, energy use, and CO2 emissions* (Special Report 298). Washington, DC: National Research Council.

Treasure Coast Regional Planning Council. (2011). *Enhancing mobility: Martin-St. Lucie 2035 regional long range transportation plan*. Retrieved from http://stlucietpo.org/pdf/rlrtp_executive_summary.pdf

Tri-County Regional Planning Commission. (2010). *Final regional 2035 transportation plan*. Retrieved from http://www.tri-co.org/trp.htm

Trost, S. G., Owen, N., Bauman, A. E., Sallis, J. F., & Brown, W. (2002). Correlates of adults' participation in physical activity: Review and update. *Medicine & Science in Sports & Exercise*, *34*(12), 1996–2001.

Tullidge, E. W. (1886). *History of Salt Lake City*. Salt Lake City, UT: Star Printing.

Turnbull, K. F., & Pratt, R. H. (2003). Transit information and promotion. In *Traveler response to transport system changes* (Transit Cooperative Research Program Report 95; Chapter 11). Washington, DC: Transportation Research Board (www.trb.org).

Turnbull, K. F., Pratt, R. H., Evans, J. E., & Levinson, H. S. (2004). Park and ride/pool. In *Traveler response to transportation system changes* (Transit Cooperative Research Program Report 95; Chapter 3). Washington, DC: Transportation Research Board.

U.S. Department of Transportation Research and Innovative Technology Administration, & John A. Volpe National Transportation Systems Center. (2012). *Best planning practices: Metropolitan transportation plans*. Retrieved from www.planning.dot.gov/documents/BestPlanningPractices_MTP.pdf

U.S. Energy Information Administration (2017). *Petroleum & other liquids*. Retrieved from www.eia.gov/dnav/pet/hist/LeafHandler.ashx?n=PET&s=MCRFPUS2&f=A

U.S. Environmental Protection Agency (U.S. EPA). (1991, January/February). *Highlights of the 1990 Clean Air Act Amendments*. Washington, DC: U.S. EPA. Retrieved from www.epa.gov/history/topics/caa90/01.htm

U.S. Environmental Protection Agency (U.S. EPA) (2001). *Our built and natural environments: A technical review of the interactions between land use, transportation, and environmental quality*. Washington, DC: U.S. EPA.

van Wee, B., Chorus, C., & Geurs, K. T. (2012). ICT and accessibility: Research synthesis and future perspectives. In K. T. Geurs, K. J. Krizek, & A. Reggiani (Eds.), *Accessibility analysis and transport planning*. Cheltenham, UK: Edward Elgar Publishing.

Verroen, E. J., & Jansen, G. (1992). Location planning for companies and public facilities: A promising policy to reduce car use. *Transportation Research Record, 1364*, 81–88.

Volpe National Transportation Systems Center. (2005). *Analysis of state long-range transportation plans*. Washington, DC: Federal Highway Administration. Retrieved from www.fhwa.dot.gov/planning/statewide/anaswplans.htm#accessibility

Vuchic, V. R., Clarke, R., & Molinero, A. M. (1981). *Timed transfer system planning, design and operation*. Washington, DC: Urban Mass Transportation Administration.

Waddell, P., Ulfarsson, G. F., Franklin, J. P., & Lobb, J. (2007, June). Incorporating land use in metropolitan transportation planning. *Transportation Research Part A: Policy and Practice, 41*(5), 382–410.

Walters, G., Ewing, R., & Schroeer, W. (2000). Adjusting computer modeling tools to capture effects of smart growth. *Transportation Research Record, 1722*, 17–26.

Wang, L., Liu, J., & Yang, H. (2015). *Transportation cost index: A comprehensive performance measure for transportation and land use systems and its application in OR, FL, and UT*. University Transportation Centers Program, USDOT. Retrieved from http://trec.pdx.edu/research/project/758

Ward, S. V. (2002). *Planning the twentieth-century city: The advanced capitalist world*. New York: Wiley.

Wasatch Front Regional Council. (2007). *Wasatch front regional transportation plan 2007–2030*. Retrieved from www.wfrc.org/publications/Adopted_2011_2040_RTP/Brochure-WFRC%20Web%20Version%2026%20Feb.%202013.pdf

Wasatch Front Regional Council. (2011). *Regional transportation plan 2011–2040: Charting our course*. Retrieved from www.wfrc.org/publications/Adopted_2011_2040_RTP/Chapter%208%20-%20Plan%20Impacts%20and%20Benefits.pdf

Wasatch Front Regional Council. (2015). *Regional transportation plan 2015–2040*. Retrieved from http://wfrc.org/VisionPlans/RegionalTransportationPlan/Adopted2015_2040Plan/FinalizePlan/RTP_2015_FINAL_UpdatedLinks2018.pdf

Watkins, K. E., Ferris, B., Borning, A., Rutherford, G. S., & Layton, D. (2011). Where Is My Bus? Impact of mobile real-time information on the perceived and actual wait time of transit riders. *Transportation Research Part A: Policy and Practice*, *45*(8), 839–848.

Weikel, D. (2015, January 13). Effectiveness of state's "parking cash-out" program is unclear. *The Los Angeles Times*. Retrieved September 16, 2015 from www.latimes.com/local/california/la-me-california-commute-20150113-story.html

Weiner, E. (1999). *Urban transportation planning in the United States: An historical overview*. Westport, CT: Praeger.

West Florida Regional Planning Council, & Florida-Alabama Transportation Planning Organization. (2010). *2035 Florida-Alabama long range transportation plan*. Retrieved from http://www.wfrpc.org/programs/fl-al-tpo/long-range-plan/2035-blueprint-documents

White, E. B. (1949). *Here is New York*. New York: Little Bookroom.

Wichita Falls Metropolitan Planning Organization. (2010). *2010-2035 metropolitan transportation plan*. Retrieved from http://wfmpo.com/uploads/plans/2/Final__1-19-10_All.pdf

Wines, M. (2016, March 28). Drilling is making Oklahoma as quake prone as California. *New York Times*. Retrieved from www.nytimes.com/2016/03/29/us/earthquake-risk-in-oklahoma-and-kansas-comparable-to-california.html

Winston, C., & Langer, A. (2004). *The effect of government highway spending on road users' congestion costs*. Washington, DC: Brookings Institution.

Winston-Salem Urban Area Metropolitan Planning Organization. (2013). *2035 transportation plan update*. Retrieved from http://www.cityofws.org/Departments/Transportation/Planning/Plans-and-Studies/2035-Transportation-Plan-Update

Yaro, R. D., & Hiss, T. (1996). *A region at risk: The third regional plan for the New York-New Jersey-Connecticut metropolitan area*. Washington, DC: Island Press.

Yoh, A. C., Haas, P. J., & Taylor, B. D. (2003). Understanding transit ridership growth: Case studies of successful transit systems in the 1990s. *Transportation Research Record*, *1835*, 111–120.

Yoh, A., Iseki, H., Smart, M., & Taylor, B. (2011). Hate to wait: Effects of wait time on public transit travelers' perceptions. *Transportation Research Record: Journal of the Transportation Research Board*, *2216*, 116–124.

Zhang, M. (2009). Bus versus rail: Meta-analysis of cost characteristics, carrying capacities, and land use impacts. *Transportation Research Record: Journal of the Transportation Research Board*, *2110*, 87–95.

Zittel, W., & Schindler, J. (2007). *Crude oil: The supply outlook*. Ottobrunn, Germany: Energy Watch Group.

Index

Note: Page numbers in italic indicate a figure, and page numbers in bold indicate a table on the corresponding page.